Do No Harm

Do No Harm

HOW A MAGIC BULLET FOR
PROSTATE CANCER BECAME
A MEDICAL QUANDARY

Stewart Justman

Ivan R. Dee
CHICAGO 2008

www.ivanrdee.com

Library of Congress Cataloging-in-Publication Data:
Justman, Stewart.
 Do no harm : how a magic bullet for prostate cancer became a medical quandary / Stewart Justman.
    p. ; cm.
 Includes bibliographical references and index.
 ISBN-13: 978-1-56663-627-8 (cloth : alk. paper)
 ISBN-10: 1-56663-627-2 (cloth : alk. paper)
 1. Finasteride—Therapeutic use—United States. 2. Prostate—Cancer—Chemoprevention—United States. 3. Prostate—Diseases—Chemoprevention—United States. I. Title. [DNLM: 1. Finasteride—adverse effects—United States. 2. Prostatic Neoplasms—prevention & control—United States. 3. Clinical Trials—United States. 4. Finasteride—therapeutic use—United States. WJ 762 J96d 2008]
 RC280.P7J87 2008
 616.99'463061—dc22   2007029867

To Linda and Marsha Frey—incomparable

I am grateful to the American Society of Clinical Oncology's 5 Alpha-Reductase Committee for including me, a layman, in its deliberation.

The opinions voiced in this book are solely my own.

Contents

Glossary

Androgen: a male sex hormone

BCPT: Breast Cancer Prevention Trial

BPH: benign prostatic hyperplasia (enlargement of the prostate)

Carcinogenic: cancer-causing

DCIS: ductal carcinoma in situ, a commonly detected form of incipient breast cancer

DES: diethylstilbestrol, a synthetic estrogen

DHT: dihydrotestosterone, a potent androgen with a suspected role in prostate cancer

Dutasteride: like finasteride, a 5 alpha-reductase inhibitor

FDA: Food and Drug Administration

Finasteride: a 5 alpha-reductase inhibitor tested in the PCPT

Five alpha-reductase inhibitor: a drug blocking the conversion of testosterone to DHT

Gleason score: a rating from 2 to 10 assigned to prostate cancer cells by pathologists depending on the cells' features, the higher scores representing the more dangerous malignancies

HGPIN: high-grade prostatic intraepithelial neoplasia, a probable precursor of prostate cancer

Iatrogenic: "doctor-caused"; used of harms

NCI: National Cancer Institute

NIH: National Institutes of Health

Placebo: an inactive substance used in the control group of an RCT

PCPT: Prostate Cancer Prevention Trial
Propecia: one milligram of finasteride, used to prevent hair loss
Proscar: five milligrams of finasteride, used to treat BPH
PSA: prostate specific antigen, a biomarker for prostate cancer
RCT: randomized clinical trial; an experimental study whose participants are assigned randomly into treated or untreated groups

I swear by Apollo the healer, by Aesculapius,
by Health and all the powers of healing,
and call to witness all the gods and goddesses
that I may keep this Oath and Promise to the best
of my ability and judgment. . . .

I will use my power to help the sick to the best
of my ability and judgment; I will abstain from
harming or wronging any man by it.

—from the Hippocratic Oath

Do No Harm

[1]

A Medical Maze:
The Finasteride Question

◦◔◦◔◦

"ALL THE grand sources of human suffering," affirmed John Stuart Mill in 1861, "are in a great degree, many of them almost entirely, conquerable by human care and effort," and among these sources he listed disease. At the time this was less a statement of fact than a proclamation of faith, though Mill's words bore fruit only four years later with the introduction of antiseptic surgery by Joseph Lister. Seemingly confirmed by one advance after another, Mill's hope has continued to animate both medical research and popular expectation. A war against cancer, inspired by the same belief in the power of "human care and effort," was declared in our own generation. But it is one thing to "conquer" cancer by timely diagnosis and effective treatment, and quite another to prevent it from entrenching itself in the human body in the first place. The second is much to be preferred. The terror that descends on a patient diagnosed with cancer, the damage wrought by such brute-force treatments as surgery and radiation, to say nothing of the still greater destruction wrought by disease that has gone beyond the possibility of treatment: all would vanish. But as devoutly as we might wish for the disappearance of cancer, a world where that affliction in all its forms has become impossible—that is, where cancer has really been prevented—is itself impossible. It is the medical equivalent of a utopian commonwealth where the very seeds of dissension and unhappiness have been plucked out of the human heart.

As used in the medical literature, the "chemoprevention" of cancer refers to the use of medication not to rule out the possibility of cancer but simply to reduce its incidence, as when women taking tamoxifen, a drug used to treat breast cancer, showed a markedly lower rate of the disease than the placebo group in the same study. Although researchers and journalists alike speak of the cancers "prevented" by tamoxifen in this case (and in deference to common practice I will use the word in the same way), it bears noting that "the term 'prevention' does not necessarily imply that the initiation of breast cancers has been prevented or that the tumors have been permanently eliminated." It suggests only that these tumors didn't appear over the period of the study, in this instance some five years. Following this study the Food and Drug Administration (FDA) approved the use of tamoxifen "specifically to *reduce the risk* of breast cancer, not to prevent it."

In 2003 a paper in the *New England Journal of Medicine* reported an advance in prostate cancer "prevention" in this qualified sense of the word. Funded by the National Cancer Institute and the National Institutes of Health, and designed to yield probative evidence, the Prostate Cancer Prevention Trial, or PCPT (I will employ this and other abbreviations for frequently used long titles), found that the steroid finasteride significantly cut the incidence of prostate cancer in men at low risk, like most who in fact develop the disease. The data on this point was so strong, mathematically speaking, that the study was brought to a halt a year early, its aim met. Finasteride thus became the first proven means to reduce the incidence of prostate cancer, and the first drug shown to reduce cancer in a low-risk group.

But although one of the study's captains in a moment of elation proclaimed the use of finasteride "the first step in conquering prostate cancer" (language that recalls Mill's prophecy of 1861), the report of the study had its dark notes. Indeed, it spoke in riddles and ambiguities, like a medical Delphic oracle. Its authors were both stirred by the promise and wary of the dangers of finasteride. No revolution in medical practice followed. As the paper in the *New England Journal* led not to a sudden and definitive paradigm shift but to other papers (it was the most commonly cited urological article of 2003), the vision of the chemoprevention of prostate cancer began to resemble a bill sent back to committee. Practicing physicians greeted the finasteride study with "tempered enthusiasm," which is to say, coolly. Despite the high hopes raised by finasteride—despite, for example, the ecstatic prophecy by an official of the National Cancer Institute that "millions of men may benefit from finasteride's ability to reduce prostate cancer risk"—urologists did not begin prescribing the drug for preventive purposes. Men remained just as exposed as before to the inscrutable dangers of prostate cancer, as women remained exposed to the terror of breast cancer despite the efficacy of tamoxifen as established in its own clinical trial. We might take the comments of a physician interviewed on CNN about the *New England Journal* paper in 2003 as exemplary of a general reaction of deflated enthusiasm. After acclaiming the paper "a very big story" and its publication "a very big day for men's health," he went on to recommend not the general use of the drug to prevent prostate cancer but consideration of its use only by men at elevated risk—that is, men whose risks were high enough to justify the use of a drug that appears, for now, to pose serious risks of its own.

In good news / bad news jokes, the bad news steals the show. So it was with the reception of the finasteride data (as also with the tamoxifen data a few years earlier): the good news of the drug's preventive effect was eclipsed by the bad news of a higher risk of aggressive malignancies associated with its use. It was as if a victory parade had been canceled. Even the press, despite its way of overplaying things and its practice of trumpeting promising medical possibilities after the scantiest research, soon gave up on the PCPT. The Olympics ban of finasteride, as a masking agent, received more play. But it may be that the real story of finasteride's clinical trial is precisely that it didn't conform to the ideal of a story—didn't provide a dramatic breakthrough. Instead of solving any of the pressing dilemmas of prostate cancer medicine, it introduced dilemmas of its own, leaving physicians with the stark question of whether or not to use this drug, the first of its kind, to prevent a disease that cries out for prevention. How physicians have answered this question reveals important things about their theory and practice of medicine.

The reasons for the general lack of jubilation that greeted the results of the clinical trial in question have to do with the bewilderingly mixed messages of the finasteride data. If the world was looking for a revolutionary finding—an advance that would mark a sharp, clean break with existing practice—the clinical trial of finasteride did not provide one. The CNN interviewer quite rightly found the finasteride data "confusing." These data I will explore shortly. For now, though, it should be noted that the kinds of vexing uncertainties surrounding finasteride are bound to reappear as other drugs with anti-cancer potential undergo clinical trial. Indeed, the finasteride question itself echoes the tamoxifen question—tamoxifen too being risky enough to be recommended for use only by patients already at elevated risk. (Tamoxifen, however, has been approved by the FDA as a reducer of cancer risk while finasteride has not.) Even if the Delphic riddle of finasteride's effects happens to be solved one day soon—a prospect called into doubt in the medical literature itself—the turn to molecular medicine that gave us finasteride in the first place will generate other questions equally puzzling. The finasteride literature introduces us to a strange, postmodern universe where cancers that do not occur can be counted, aggressive malignancies can be discounted, the same drug can exert seemingly contrary effects, and incalculables are factored into rational models. If cancer prevention in general is heading in this direction, perhaps the experience gained in the finasteride controversy will enable us to go about prevention wisely. I say wisely because despite the finasteride literature's surrealism, most of its conclusions are cautious, as if the authors were mindful after all of the ancient maxim and framing principle of medical practice, Do No Harm. One researcher's review of the potential shown by certain drugs like tamoxifen as prostate-cancer preventives concludes that in order to go into use, they would have to be "very safe and have no genotoxic [that is, DNA-destroying] or other adverse properties." This imperative, while usually unstated, also governs the finasteride debate.

The same researcher wrote in 1992, just before the inception of the PCPT, that "some of the most promising agents in the chemoprevention of prostatic cancer," among them the so-called 5 alpha-reductase inhibitors (of which finasteride is one), "may enhance rather than slow down the progression of prostatic cancer." (Five alpha-reductase inhibitors block the conversion of testosterone into the more potent androgen, dihydrotestosterone.) The PCPT data suggest that finasteride may do both at once. Seemingly capable of suppressing prostate cancer (a disease that, as of January 2003, had touched two million living American men) as well as promoting it, this double-edged drug, finasteride, tests our principles and compels us to state them. Do we believe that a reduction in the general incidence of prostate cancer justifies an increase in the most dangerous malignancies? Should we ask a drug to promote the greatest happiness of the greatest number, as if it were a public policy? Should finasteride go into general use as a cancer preventive while much remains unknown about its workings and effects? To this point, medicine has returned the answer, No, to each question—a restraint all the more remarkable in view of the generally activist approach to prostate cancer in American, as opposed to European, medicine. One might have expected American urology, which screens for prostate cancer more aggressively than its European counterpart (and treats detected disease accordingly), to take a similarly aggressive approach toward chemoprevention. After all, some would say that prevention is simply early detection taken to the next level. Indeed, there is stronger evidence that finasteride reduces the incidence of prostate cancer than there is that PSA screening, a blood test widely used to detect the disease, reduces mortality as a result of prostate cancer. Finasteride's evidence arises from a randomized clinical trial (the acid test of medical efficacy), while such trials of PSA are still awaited.

In a revised version of the Hippocratic Oath now in use in many American medical schools, the initiate pledges to "prevent disease wherever I can, for prevention is preferable to cure," and to avoid "those twin traps of overtreatment and therapeutic nihilism" (that is, suspicion of excessive and uncertain treatments). Amid its many fine sentiments, the oath conspicuously does not, however, include a promise not to harm. Together these features would virtually dictate the use of finasteride, a drug proven to prevent prostate cancer, and to date the only way out of the miserable impasse of overtreating the disease with surgery and radiation. And yet doctors—undoubtedly including any number who have taken this very oath—do not prescribe finasteride for prevention because they are deterred by its risks, which is to say, restrained by the duty not to harm which the revised oath discreetly sweeps under the rug. In the same way, the finasteride papers recite the benefits of prevention and the costs of overtreatment, and proceed to establish that the drug does in fact cut the incidence of prostate cancer markedly, only to refrain in most cases from actually endorsing it because it appears finally too risky. Although quietly edited out of the new version of the Hippocratic Oath, the duty not to harm still exerts its power. Thus while the new oath commits the physician to guard against the passivity of "therapeutic nihilism," most urologists have chosen to do nothing in response to the dramatic data of the PCPT. Even in this postmodern era, and in spite of its own activist leanings, American urology remains in touch with established principles of caution. If the medical literature on finasteride is a labyrinth, the rule Do No Harm is the thread.

According to the philosophical tradition underlying the practice of modern medicine—the tradition of empiricism—knowledge is founded on observation and verified by experience. When the PCPT began in 1993, experience with finasteride was limited, the drug having just come into use, and this being so, perhaps it is not a mystery that even after the PCPT, knowledge of its effects remains limited. (Even in the case of tamoxifen, which unlike finasteride had a history of use, and as a cancer treatment at that, before being tested as a chemopreventive, no one knows what the effects of long-term use will be. Experience is lacking.) The directors of the PCPT originally envisioned a clinical trial that would measure the different survival rates of the participants, but when it became clear that such an experiment would require some fifty thousand volunteers and extend over fifteen years, it was rejected as hopelessly unwieldy. Instead they devised a trial to measure the different incidence of prostate cancer in the treated and untreated groups, even though the incidence of prostate cancer does not reliably predict mortality. The paradoxes and puzzles of the PCPT thus began at conception, designed into a study in chemoprevention that could not really say much about the goal of prevention itself: survival. The PCPT was considered concluded, and its results made known, when the reduced rate of prostate cancer in the finasteride group was established, not when the survival rates of the two groups were determined.

In a sense, then, the finasteride question is a question because as a world we are in a hurry. Physicians who, despite the hurry, have decided to wait and see before recommending finasteride for cancer prevention are squarely in the honorable tradition of empiricism, which in England traces back to John Locke, himself a physician who believed in caution and distrusted dangerous (or as they came to be called, heroic) measures as well as being a friend and admirer of "the English Hippocrates," Thomas Sydenham (1624–1689).

Referring as it does to the practice of ousting disease from the human frame by some force as powerful as disease itself—for example, violent purging and serial bloodletting—"heroic" medicine almost necessarily entails harm. Because traditions do not expire like snuffed candles, the tradition of heroic medicine persisted even as medicine modernized. Not for a century and more after Locke's death was the practice of medicine placed on a scientific foundation, and when it was, it stood against violent approaches like overdosing on the one hand and passive approaches like trusting to Nature on the other as harmful, both of them, to the patient. *The emergence of medicine as a profession and a science in the nineteenth century coincides with the reaffirmation of the Hippocratic principle of not doing harm.* Only then did medicine begin to resemble the practice we know today. Only when doctors became aware that they themselves were the transmitters of infection—when they recognized their own "irreparable errors and wrongs," in the words of Oliver Wendell Holmes—only then was the terror of puerperal, or childbed, fever reined in. Only when hospitals themselves ceased to be houses of infection, places whose very name connoted pestilence, did they become something we would recognize as medical institutions; and this was accomplished in the nineteenth century. All of this bespeaks a new looking-inward and a new wariness of causing harm.

In 1849, two years after the founding of the American Medical Association, the prominent physician Worthington Hooker began his exposé of medical errors and abuses by citing "the multiform and often fatal injuries" caused by irresponsible practitioners. Simultaneously doctors began to deplore the loss of even a single patient to chloroform—perhaps the first time the rule Do No Harm had been applied so stringently. At this stage of medicine, even if doctors could rarely cure their patients, as Edward Shorter notes, "at least an understanding of disease mechanisms and drug action *kept them from doing harm* [emphasis in the original]. This ability to refrain from doing harm stands as one of the major acquisitions of primary care for the period from around 1840, when bloodletting began to go out of use, to 1935, when the first of the wonder drugs was introduced." Now dismissed by critics as a mere piety, the Hippocratic rule in the nineteenth century was strong medicine—a bill of indictment, a summons to responsibility, an agent in the transformation of medicine into the institution that now takes its own origins for granted. It is a striking irony that our most venerated profession began to acquire its identity during a period now popularly detested as one of stifling repression—the Victorian era. For all our alienation from tradition, for all our image of medical progress as one breakthrough after another, medicine as we know it is a child of the past.

The idea that the past has anything of value to contribute to the present may have an unfamiliar sound in an age that tirelessly celebrates its emancipation from tradition. So broad is the bias against tradition that even in the humanities, those fields of learning most concerned with historical records, times past are sometimes portrayed as an era of darkness, and persons past as prisoners of their era, unlike ourselves. The medicine of the past shares in the disrepute of the past in general. We hear of the Tuskegee Syphilis Experiment, in which hundreds of infected black men were deliberately kept untreated for decades, even after the advent of penicillin, in order that science might observe the ravages of the disease. (I discuss this infamy later.) We hear that doctors lied to patients freely, as in the Tuskegee case, and mutilated women with impunity. We hear less about physicians who experimented on themselves or raised the standards of medicine. My sense is that the depiction of the past as a museum of horrors is a caricature: it contains an element of truth, but truth distorted through a lens of bias and animus. If, as seems undeniable, physicians once lied to patients more readily than they do now, one reason, aside from the absence of lawsuits to keep them honest, must have been that some sincerely believed the truth can harm.

Whatever we may think of paternalistic lies, the principle of not doing harm is certainly an estimable one. This principle could have been indelibly tainted by its association with the old ways of paternalism, and could have been discarded some decades ago when the entire edifice of paternalism fell into disrepute, but it was not. That it persists at some level to this day, despite its tattered and sullied history, and in the face of our own suspicion of the past, suggests both the power and the depth of tradition. Of William Osler (1849–1919), a founder of the Johns Hopkins Medical School and a uniquely revered figure in modern medicine, it is said that his writings "everywhere evince an appreciation of the Hippocratic obligation 'to be of benefit and do no harm.'" Osler's caution was well founded. He was born at a time when "the fatal termination of many illnesses was attributable to the doctor rather than to the disease," before antiseptic surgery, in an era when drugs were used haphazardly and the practice of medicine, despite the advances of the nineteenth century, was still colored by theories as unscientific as astrology. A favorite drug was mercury.

According to the story line of medical history, the advent of antibiotics in the 1940s wrote an end to the era of therapeutic nihilism, when professional medicine viewed remedies suspiciously in the belief that little could be done with most diseases beyond letting them cure themselves. In its time, though, therapeutic nihilism was a reaction against the practice of harming the patient. To the extent that "the fatal termination of many illnesses" was the work of the doctor himself, those who treated the patient conservatively, guided by the principle of at least not doing injury, advanced the practice of medicine. Compared with penicillin, therapeutic nihilism may appear defeatist and backward, but "defeatism" is surely a prejudicial description of the honesty it took to recognize that medicine itself was responsible for great harm, and specifically that "the stupendous therapeutic armamentarium of the [nineteenth century], which included everything from botanical extracts to applications of electrical currents and leeches, simply did not work." Without the bold skepticism of our forebears, those electrical currents and leeches might still be in use (demand for nostrums runs high to this day, after all); and both skepticism and intellectual honesty remain indispensable in medicine, all the more so now that, in the words of Lewis Thomas, "the stakes have become very much higher, and the possible remedies for disease much more powerful and potentially dangerous," as in the case of finasteride.

There is no denying that folk practices such as bleeding and purging, intended to rid the body of what ailed it, harmed. As for traditional diagnostic techniques such as feeling the pulse, listening to the breath, and inspecting or tasting urine, they did not take medicine very far, though at least none of them injured the patient. That like bleeding and purging they presume a patient, a single person, bears stating because it is from the past, too, that we inherit the principle that medicine centers on a doctor and a patient—an axiom we may take for granted because it is embedded so deeply in our thought and practice, but which not everyone in fact grants. After declaring the rule Do No Harm impractical, a physician writing in the *British Medical Journal* goes on to argue that

> The single greatest difference between Hippocratic medicine and modern medicine is that the former focused solely on the relationship between the individual doctor and the individual patient, whereas the latter also takes the concerns of society at large into consideration. The emphasis is shifting increasingly toward

the concerns of society. This shift, which began 200 years ago with the issue of public health, is utilitarian. Aiming to achieve the maximum benefit for the greatest number has necessarily interfered with the Hippocratic ideal of medicine.

In the estimation of this critic, the principle that medicine "centres solely on the individual" is both obsolete and injurious to the public welfare.

If in some cultures "villagers cluster around a healer and a patient" like spectators of a drama, we can perhaps understand what it is about the relation between one patient and one doctor that would offend a resolute modernist like the anti-Hippocratic physician. It is too ancient. Just as humanity had to learn that the earth was not the center of the universe, so (it is implied) patients will have to learn that they are not the center of the medical universe. Already insurers are third parties to the doctor-patient relationship. It is true too, of course, that a physician treating a single patient with a drug of unknown efficacy is in no position to determine the drug's effect; for that, only numbers of patients under controlled conditions will do. Still, if you are a breast cancer patient confronting the delicate decision of whether or not to take tamoxifen, you might well hope to talk things over with a physician willing and able to concentrate on your case. "It must be kept in mind that for each patient the benefits from . . . tamoxifen therapy must be weighed against the risks and costs." For each patient. To define the welfare of society in general, not the patient, as the highest good of medicine is to fall in with powerful trends in the social sciences and even in the humanities that submerge the person in classes or statistical categories.

These abstract considerations bear directly on the finasteride controversy. It is highly probable that at this moment someone really committed to "the maximum benefit for the greatest number" would endorse and encourage the use of a drug capable of cutting the incidence of prostate cancer significantly, albeit at the risk of some aggressive malignancies here and there. Indeed, the argument is already being made in print that, regardless of this risk, it makes sense to use finasteride broadly for prevention because it suppresses so much cancer. If most practicing physicians do not take this position, it is because they are deterred by the tradition of Hippocratic medicine that some, like the cited critic, deem both unworkable and archaic.

The finasteride controversy is important not only in itself and as an omen of the future conundrums of chemoprevention but as an instance and marker of the trend toward population-based medicine. The cited critic of the Hippocratic code is not alone in his belief that medicine no longer centers on the physician-patient relationship, for as an observer wrote in 1990, "The new medicine just coming into being differs from current medicine in that its science is population based rather than patient based." In this spirit, some of the finasteride authors project the statistical benefits to be reaped if the drug were used preventively "at the population level," even assuming the associated risk of high-grade cancers to be double what it now appears. These theorists of a new medicine are able to contemplate the doubling of a risk of aggressive cancer that is quite enough as it is to deter physicians on the ground from prescribing finasteride preventively. The principle of not doing harm does not speak to the physician doing medicine at the population level as it does to the physician whose patients have bodies.

Some might point out that the text of Hippocrates' *Epidemics* admonishes physicians "to help, or at least to do no harm," as if to say, "Considering how little we physicians can actually do for our clients, do not make matters even worse by harming them. Let the body heal itself." On this showing, the celebrated first principle of medicine merely reflects the impotence of the medical art at the time it was framed, and now that medicine has grown immeasurably more powerful—now that it can do a great deal for patients—that cautionary maxim can be put to one side.

After all, medical tools powerful enough to do much good will always entail the risk or element of harm. Chemotherapy kills healthy cells along with cancerous ones. Merely admitting patients to a hospital "exposes them to unusual and virulent pathogens. Even simple bed rest predisposes to thromboembolic disease." For all that, it remains true that some preventive measures of the first importance, such as not smoking—and the finasteride debate concerns prevention—in fact cause no harm. Nor is it the case that medicine has progressed to the point that doctors no longer need to be constrained by the principle of not doing harm. Both tamoxifen and finasteride are in use in spite of not much being known about how they work. Neither drug was originally synthesized as an anti-cancer agent; their use for that purpose is empirical. Appropriately for risky drugs not well understood, they are not employed across the board for cancer prevention. That neither is so used despite fulfilling projections in their clinical trials, and despite the sheer urgency of the universal desire to prevent cancer, testifies to the endurance of the medical principle of avoiding harm.

Regardless of the categorical prohibition of harm, medicine cannot afford to avoid all harms, only harms not justified by benefits. So routine is the weighing of harms against benefits that the balance scale has become, in effect, the emblem of medicine as well as of justice. But what is meant by balancing harms and benefits? If a certain vaccine prevented more cases of disease than it caused, its benefits might outweigh its harms in some bare, minimal sense, and arguably it would serve the greater good, but this doesn't mean its use would be medically defensible. In order to justify use of the vaccine we would require it to prevent not just more but overwhelmingly more casualties than it causes. Precisely because it is medical in nature, the vaccine is held to a higher standard than another venture might be. In the 1954 trial of the Salk polio vaccine, among 400,000 children inoculated, 204 cases of vaccine-associated polio, leading to 11 deaths, occurred as a result of a defective batch of vaccine. Such was the shock of these numbers that congressional hearings were held.

Consider an example closer to the matter of cancer prevention. It is reported that

> The randomized trial showing the largest benefit of breast cancer screening . . . found a difference of 22 breast cancer deaths (22 fewer among women screened than among those not screened) among a total of 1,000 deaths overall. . . . All it would take is a few deaths somehow related to screening and the positive effects would diminish, or even disappear.

Note that the author doesn't say there would have to be twenty-three screening-related deaths in order to overbalance the benefit of lives saved by screening; in fact he doesn't state a number at all. "All it would take is a few deaths." This vagueness seems just. The balance scale is not to be taken too literally. Morally speaking, three or four deaths might overbalance twenty-two if these three or four deaths came about at the hands of the medical profession itself, charged as it is not to do harm. There is something about the close calculation of a break-even point that doesn't really befit the ethos of medicine. When certain contributors to the finasteride debate compute exactly what the risks of the drug would have to be to cancel out its survival benefits in person-years, they cease to think like doctors and begin to think like actuaries, statisticians, or investors. The model of the balance-scale weighing risks and benefits cannot really accommodate a question as fraught with imponderables as the finasteride issue.

In order for two entities to be weighed in a balance scale they must be commensurable. When it became clear that the arthritis medication Vioxx, then still on the market, was associated with a significantly higher rate of heart attacks, patients and physicians were left to weigh the benefit of relieving arthritis pain (for which, after all, Vioxx was not the only means available) against an increased risk of the leading cause of death in the United

States. Benefit and risk are simply not comparable. The difference between low- and high-grade prostate cancer is not as categorical as that between arthritis pain and a heart attack, but it *is* strong enough to throw into question the very commensurability of the risks and benefits of a preventive regimen of finasteride. The effect of all the talk of "balancing" risks and benefits in the finasteride papers is to normalize a degree of danger that only recently would have been unthinkable in preventive medicine. Given current knowledge, the preventive use of finasteride would not only put untold numbers of men at risk of high-grade cancer but would undermine standards of safety themselves. Perhaps a reluctance to set such an ominous precedent has contributed to the profession's unwillingness to see the drug into general use.

The balance scale is such an appealing image of just deliberation (indeed, the word "deliberation" derives from "libra," meaning scale) that we can be misled into assuming that it can decide the most haunting questions. It is well known by now that physicians performed sadistic medical experiments in Nazi death camps. Assuming that the results of their investigations are not junk science but have some technical value, the question arises, Should researchers now use them? Some take the position that

> if the benefit of the research is of such sufficient magnitude that it saves more lives than were lost acquiring the data, and the data could not or would not have been obtained in any other way without entailing such human suffering, then perhaps we should acknowledge and use the data and at the same time express the highest censorship and opprobrium to the investigator [sic] each time the data are discussed.

But even if we knew exactly how many lives were lost in the Nazi experiments, how is it possible to know how many victims of, say, hypothermia stand to be saved over the years to come? In this case, are we not relying on the familiarity of the balance scale to defuse the shock of the question itself, and to decide an issue that should not and cannot be decided by comparing quantities? To my mind, the question of what to do with the findings of Nazi medical experiments is best disposed of with a summary judgment: the findings represent so much forbidden knowledge. (As it happens, the *New England Journal of Medicine* refused to publish the findings of the Nazis' hypothermia experiments.) Note, though, that the Hippocratic Oath "certainly was recognized in Germany" at the time of these sadistic exercises. If the responsible physicians and their many complicit brethren had not convinced themselves that their duty to the health of the Aryan body politic overrode the recognized duty not to harm, the extermination of European Jewry would have lost its medical pretext.

Critics of Hippocratic medicine might reply that that was then and this is now, and that at this moment the principle Do No Harm resembles an honorific motto like *Lux et Veritas* emblazoned over the entrance to a library—uplifting perhaps, but of no effect. They might remind us that this august principle was born in backwardness, and that the oath associated with it "does not mention informed consent and disavows surgery." They might observe that one can avoid doing harm by doing precisely nothing, which is a very poor sort of medicine; or that placebos do no harm. They might argue that the principle of avoiding harm is so inimical to the advance of medicine that for two millennia it inhibited the dissection of the human body, depriving the practitioners of medicine of the most necessary knowledge. They might note that under the Hammurabi Code, "if a free citizen died while in the physician's care, the physician's hand or fingers were to be cut off," and that it is time, by now, to let go of the obsession with blaming doctors for the harms wrought by medical care. With all that said, if finasteride were widely prescribed for prevention, only to induce aggressive prostate cancer "at the population level," future historians would not hesitate to condemn all the physicians involved for losing touch with the axiom Do No

Harm. But as things stand, physicians are not about to prescribe finasteride for prevention. After the worst has been said of the principle of not doing harm, and of tradition itself, they are still not of a mind to do away with the traditional guiding precept of their profession.

The Hippocratic rule after all has a great deal built into it. It belongs to the body of belief that diseases are physical in nature, the originating belief of Western medicine (reflected in the very word "physician"). Insofar as it may once have distinguished the Hippocratic physician from the miscellaneous practitioners who did cause harm, it underwrites the medical profession's sense of itself *as* a profession—a corporate entity with a set of standards higher than those of the marketplace in general. It was when Sydenham, citing Hippocrates, discarded dogma and speculation in favor of precise clinical observation, and exercised a principled caution in the use of medicines (for "nature alone often terminates diseases, and works a cure with a few simple medicines, and often enough with no medicine at all"), that English medicine found the modern track. It was when physicians acquired the knowledge of what they could not do—the knowledge that *"kept them from doing harm"*—that primary care, and indeed medicine generally, assumed the character by which we now know them.

Transforming events in the history of medicine include the introduction of chloroform anesthesia by James Simpson, the introduction of antisepsis by Lister (with the consequent reduction of high rates of fatal infection at the hands of medicine itself), and Paul Ehrlich's synthesis of a drug, Salvarsan, to cure syphilis. Simpson administered chloroform to thousands of patients without a death. Lister and Ehrlich sought a way to kill a pathogen (the sources of hospital infection, and the bacterium responsible for syphilis) without injury to the patient, and both spoke specifically of doing so. All three seem to have been not just constrained but inspired by the principle Do No Harm. To repudiate the Hippocratic rule as an archaism is also to dismiss much that defines medicine as a science, a profession, and a moral practice, which not many are prepared to do.

Whatever their political differences, most Americans hold to the principle that we should be free to do as we like in our private affairs, provided we cause no harm to others. By far the most influential defense of this position is laid out in John Stuart Mill's essay *On Liberty*, published in 1859, four years after John Snow's now-classic study tracing the spread of cholera in London to an infected water supply, two years after the inauguration of the *British Medical Journal*, one year after the Medical Act establishing a registry of properly qualified physicians, the same year as Florence Nightingale's *Notes on Nursing*, one year before the phrase "primum non nocere"—"first do no harm"—was coined, two years before Prince Albert, aged forty-two, died of typhoid fever, five years before Pasteur's germ theory of disease, and six years before Lister's invention of antiseptic surgery, the advance that "marks the moment at which doctors began to be able to save lives." In defending freedom of thought and discussion, Mill makes the argument that ideas that were once alive decline over time into shadows of themselves unless reanimated by the spirit of contention. This, he says, is the fate of "almost all ethical doctrines and religious creeds": "full of meaning and vitality to those who originate them," they sink gradually into the category of received ideas, until finally "instead of a vivid conception and a living belief there remain only a few phrases retained by rote," espoused but not acted upon. Only by controversy, Mill believes, can such doctrines be recalled to life. Around the time Mill wrote these words, the ethical doctrine governing medicine, Do No Harm, was itself being renewed and reaffirmed, converted from a classical tag to a living practice, awakened into urgency by the controversies swirling around anesthesia and antisepsis, and by the growing recognition that unconscionable harm was being done in the name of medicine. To the revolution that ensued the medicine of today can be traced.

Along with freedom of opinion, Mill defends freedom of action within certain definite limits. There exists, says Mill,

> a sphere of action in which society, as distinguished from the individual, has, if any, only an indirect interest; comprehending all that portion of a person's life and conduct which affects only himself, or if it also affects others, only with their free, voluntary, and undeceived consent and participation.

But what of dueling? After all, the parties enter into a duel in an explicitly consensual manner, with eyes wide open. Although dueling figures significantly in the literature and lore of the nineteenth century (suggesting that the issue was still alive), Mill does not even mention the practice, as if it simply went without saying that he could not approve such a barbarity, whether or not it technically conforms to his definition of an act outside society's jurisdiction. The prohibition of harm trumps even the principle of free and undeceived consent. In other words, Mill implicitly counts on the traditional ban on killing to rule out dueling, even if the tendency of his polemic might seem to leave an opening for it. Despite his dislike of mere tradition, Mill relies to this extent on tradition to control the libertarian implications of his argument. Let this serve as a rough analogy to the restraining effect of the medical principle Do No Harm in the finasteride papers. As the proscription of killing implicitly qualifies Mill's libertarian argument, so does the proscription of harm implicitly frame and govern the finasteride debate, tempering its conclusions. However indifferent to the ways of the past the participants in the debate may imagine themselves, they are not about to shelve the rule Do No Harm—and a good thing, too.

*

According to one version of events, it was the titan Prometheus who introduced humankind to medicine. "For lack of drugs they wasted until I showed them blendings of mild simples with which they drive away all kinds of sickness." Inasmuch as "Prometheus" means "forethought," it seems that some notion of anticipatory action, of prevention, is built into the mythic origins of Western medicine. Today planning enthusiasts use the neologism "proactive" to describe the sort of attacking defense that keeps problems from arising, as opposed to merely reacting once they are already upon us. Cancer prevention is in this spirit. It recalls medicine's origins in forethought even as it raises visions of the future. Working at the very boundaries of knowledge, chemoprevention marks one of the frontiers of medical theory and practice, though being a frontier—that is, more or less unmapped territory—the field is fraught with uncertainty as well as promise. The finasteride question crystallizes this ambiguity.

As I write, no one knows how the finasteride question will be resolved. The book is still open. While we now know who was on the right track in early trials of a polio vaccine, and that the man who sold the patent to aspirin to the Bayer Company got shortchanged, and that the opponents of anesthesia were on the wrong side of history, this sort of knowledge, possible only after the fact, is not yet available regarding finasteride. The superior vantage of retrospect denied us, we find ourselves in the same position of not knowing as those who went before. Indeed, our predecessors in ignorance have things to teach us. In the eighteenth-century controversy over smallpox inoculation, one party denied "that if inoculation only caused one victim out of ten to perish, it would still be advantageous for the sole reason that it would increase the average life by some days." Analogous issues and arguments appear in the finasteride papers. In *Madame Bovary* (1857) a high-and-mighty surgeon disbelieves in anesthesia and quite erroneously dismisses the possibility that a club foot can be corrected; but his principle of never operating on a healthy patient is no foolish dogma. Many European urologists oppose screening for prostate cancer because it leads too often to surgery that confers questionable benefit but carries the risk of impotence and incontinence. Some would say that administering a drug like finasteride to a population of men resembles operating on a healthy patient.

If the end of the finasteride story has not yet been written, it begins, like many a good tale, a long way from the course it later took. Thirty years ago it was discovered that a cluster of male pseudo-hermaphrodites in the Dominican Republic—men born with ambiguous genitalia, and as children resembling girls—inherit a deficiency in an enzyme governing the conversion of testosterone into a more potent form. As adults they have an undeveloped prostate—and are also exempt from prostate cancer, placing them among the few of whom it can truly be said that the disease has been ruled out. Given the high proportion of aging men in the United States who suffer from enlargement of the prostate (a condition then treated with surgery), researchers began to wonder if the same enzyme deficiency might not be induced in these patients with beneficial results. Finasteride was born. First given experimentally to human subjects in 1986, the drug was on the market as a treatment for enlarged prostate a few years later, less than two decades after the original papers on pseudo-hermaphroditism. That is the most straightforward part of the finasteride story. From there it spirals into perplexity.

[2]

The Prostate Cancer
Prevention Trial

◦◖◗◦◖◗◦

I. THE PROSPECT OF PREVENTION

It is estimated that one in six American men will be diagnosed with prostate cancer. The fraction of the male population who will die of the disease is far smaller—one in thirty—which implies that most such cancers are not life-threatening and that, consequently, much of the treatment of the disease represents overtreatment. With unintentional irony, researchers commonly refer to the "risk of diagnosis of prostate cancer" instead of the risk of the disease itself. A provocative essay in the *New York Times* recently cited the anomaly of prostate cancer diagnosed "in over a million people who, but for testing [that is, screening], would have lived as long without being a cancer patient." But considering that only a few decades ago people lived no more than a year or so once cancer became apparent, some effort to identify and treat the disease in its early stages was surely called for. If prostate cancer were detected only by the unreliable and rather primitive method of the digital rectal exam—that is, if the blood test responsible for the massively increased detection of the disease since the later 1980s were not in use—cancer would already have escaped the prostate, thus becoming incurable, in fully half the cases of discovered disease. With respect to prostate cancer, the stark disparity between incidence and death rate, far from being cause for comfort, merely points to the elusive, poorly understood nature of a disease that varies from the "clinically insignificant" to the lethal. It seems that prostate cancer itself speaks in ambiguities (as could be said of breast cancer as well).

If five times as many men will be diagnosed with prostate cancer as will die of it, this still means that, numerically, the disease is worse than Russian roulette. As with that pastime, the prostate cancer patient is at the mercy of the unknown, because there are no markers of any kind to distinguish a lethal malignancy of the prostate from a nonlethal one. "Either because of or in spite of this uncertainty, in the US 95% of patients with newly diagnosed [prostate tumors of the most common grade] choose some form of definitive therapy instead of watchful waiting." In view of the dangers of leaving prostate cancer untreated, the unwelcome side effects of even successful treatment, the harms of overtreatment (and "there is probably no cancer in which the existence of pseudodisease is more widely accepted and the value of early detection more vigorously questioned" than prostate cancer), and every prospect of that much more of all these woes as the American population ages, the ideal of preventing the disease from appearing in the first place is all the more compelling. The long latency period of prostate cancer—some twenty to forty years—makes the disease an especially good candidate for prevention. And as drugs known as 5 alpha-reductase inhibitors came onto the market and into use in the early 1990s, the possibility of prevention began to suggest itself. Five alpha-reductase inhibitors like finasteride block the conversion of testosterone into the far more potent dihydrotestosterone (DHT), an androgen implicated both in enlargement of the prostate (BPH) and prostate cancer. Finasteride shrinks the prostate, thereby treating BPH—so successfully, in fact, that some waggish souls predicted that before long, so many men would be taking the pill that candidates for the placebo side of a prevention trial "would be difficult to find." In view of finasteride's dramatic performance as a therapy for BPH, researchers began to wonder whether it might not shrink the incidence of prostate cancer as well. It was to test this theory that the Prostate Cancer Prevention Trial (PCPT) was conducted.

At the time, the PCPT was the largest urological study ever carried out. Begun in 1993, before the supply of men not yet being treated for BPH ran dry, the experiment randomly divided 18,882 subjects—55 years of age and older, but at comparatively low risk for prostate cancer—into two groups, one of which received a placebo, the other five milligrams per day of finasteride. Neither the patients nor their doctors knew to which side of the study they had been assigned. Both groups were kept under regular medical surveillance, biopsies being performed if elevated PSA was detected or digital exams revealed abnormality. After seven years, beginning in 2000, men who had not yet been biopsied were "offered" an end-of-study prostate biopsy. (Both PSA and digital exams are rough measures, at most pointing to a need to biopsy. Only with the examination of tissue specimens by pathologists is the presence of cancer confirmed. At the study's peak, FedEx trucks carried such a cargo of the miniscule specimens to the central pathology laboratory in Denver that the facility was overwhelmed.) Over the years of the study the participants dwindled by attrition, the final figure still being large enough, however, for statistically robust results. The men and their medical fortunes are still being tracked.

By the end of the experiment, prostate cancer was found in 803 of the 4,368 men in the finasteride group (18.4 percent) and, 1147 of the 4,692 in the placebo group (24.4 percent), meaning that the cancer rate in the first group was almost 25 percent lower. (Considering these numbers, it is remarkable that only five men in each group died of prostate cancer over the course of the study.) Evidently finasteride does shrink the probability of prostate cancer, just as investigators theorized. A notable irony stares out at us in the numbers, however. Even the reduced cancer rate of the finasteride group (18.4 percent) exceeds the *lifetime* risk of prostate cancer of men in the United States (17 percent). Why are the cancer rates of both groups in the PCPT so high? Presumably because the surveillance regime of the study itself brought to light cancer that would ordinarily have escaped medical notice. The question of exactly how many men have prostate cancer "depends on how hard you look," and the PCPT looked microscopically. About half of all cancers detected in the PCPT came to light in end-of-study biopsies of men in both the treated and untreated groups who showed no signs of the disease whatever, just as unsuspected prostate cancer shows up commonly in autopsies. Men in the experiment were indeed at increased risk of diagnosis, and for those on the finasteride side the statistical benefits of the drug were offset by the study itself. While lost in the glare of more dramatic findings, the high frequency of cancer on both sides of the study, but especially in the control group, is surely one of the stories of the PCPT. How many of those cancers discovered only after the most assiduous searching were threatening? What does it mean to say that prostate cancer stalks fully a quarter of the male population of the United States?

Some numbers are stories in their own right, and so it is too with a figure we began with, the lifetime risk of death as a result of prostate cancer in the American population. That number is actually two numbers: 2.8 percent for whites, 4.7 percent for blacks. One of the unsung findings of the PCPT is that finasteride's preventive effect cut across all divides, including those of age and race. In principle, therefore, finasteride could do something about the alarmingly high rate of prostate cancer, and death by prostate cancer, in black Americans—numbers that need all the reduction they can get. Of the black men in the control group of the PCPT, fully 34 percent were found to have prostate cancer.

2. THE RUB

The PCPT, then, demonstrated that a pill could reduce the incidence of prostate cancer by a quarter. It is true that finasteride has sexual side effects that cannot be pretended away (of this more as we go), but even so, it would almost certainly be in use at some level for preventive purposes if the 25 percent reduction were the end of the matter. But it is not the end of the matter. If the finasteride group in the PCPT showed a reduced incidence of prostate cancer that nevertheless exceeded the incidence in the general population, this could not be blamed on finasteride itself. But the finasteride group also showed a markedly higher rate of more aggressive, or high-grade, tumors. For all biopsies performed, 37 percent of the detected cancers in the finasteride group (280 of 757) but only 22.2 percent of those in the control group (237 of 1,068) proved to be high-grade. Thus the same drug that lowered the general incidence of prostate cancer was also associated, for some reason, with a disturbing increase in the most dangerous forms of the disease. Observe that not only the percentage but the absolute number of high-grade tumors was higher in the finasteride group.

Let it be understood at the outset that for one substance to inhibit and induce cancer is not biologically impossible. "Many chemopreventive agents" exert "multiple effects, both positive and negative." Not only does finasteride appear to fall into this paradoxical category, but no one can rule out the perverse possibility that in the PCPT it suppressed just those cancers *least* likely to prove deadly, much as PSA screening has detected legions of malignancies but as yet has not made much of a dent in prostate cancer's death rate. For the medical community as well as the press, in any case, the increase in high-grade disease cast a pall over finasteride's otherwise inspiring showing in the PCPT. "Ay, there's the rub." Notice too that while the PCPT authors report the lower rate of cancer in the finasteride group as a percentage (24.8

percent), they do not do the same for the higher rate of specifically high-grade cancers. (Other finasteride papers follow the
same procedure, as in this, the first sentence of an article on the
PCPT: "In the Prostate Cancer Prevention Trial, men receiving finasteride had a 24.8 percent lower risk of prostate cancer than
men receiving placebo but a higher risk of high-grade cancer.") If
both benefits and risks were reported as percentages, surely it
would look as if the second offset or overshadowed the first.

The increase in high-grade cancer among the finasteride
group is alarming because, in addition to its other ironies,
prostate cancer behaves virtually like two diseases. Tumors of the
prostate are graded according to their microscopic features on
what is called the Gleason scale, from two to ten, with most cancers graded six and below responding well to treatment, and
those seven and above proving more dangerous and intractable.
Men with the highest grades of prostate cancer stand a twenty
times worse chance of dying within ten years of diagnosis than
men with the lowest grades have of dying within twenty. In the
study that established the Gleason system itself, a graph of mortality rates by Gleason score looks like a series of rather steeply
ascending steps. (So serious a view is taken of high-grade cancer
generally that when Congress passed legislation to regulate medical laboratories in 1988, the Centers for Disease Control issued
a rule that any person would fail its accreditation test "if he or she
interprets as negative or benign even one slide that shows a high-
grade lesion or cancer.") Gleason grading is not an exact science,
however, nor is the dividing line of Gleason six absolute. Many
specimens graded six prove deadly or are later upgraded, while
conversely some high-grade malignancies do not. In the subtle,
mysterious, elusive disease of prostate cancer, "while tumor grade
correlates with outcome, some high grade tumors never lead to
mortality while a third of deaths from prostate cancer occur in
men with Gleason 6 or less disease." Nevertheless it remains generally true that the higher the Gleason score, the more dangerous
the case.

The following table records some of the most pertinent PCPT data:

PROSTATE CANCER PREVENTION TRIAL
Finasteride vs. Placebo

	Finasteride	Placebo
Prostate Cancer	18.4%	24.4%
BPH	5.2%	8.7%
Gleason 7-10	6.4%	5.1%
Loss of Libido	65.4%	59.6%

Sources: Barnett Kramer, MD, MPH
Ian Thompson et al., *New England Journal of Medicine*, July 17, 2003

Of the cancers detected in the PCPT, most by far were assigned Gleason scores of six, but distinctly higher rates of Gleason seven through ten, and especially eight through ten, appeared in the finasteride group, in biopsies performed both for cause during the study and for purely investigative purposes at the study's end. It bears emphasizing that higher scores represent the more malignant and aggressive cancers, those that impose the heaviest burden of suffering. Among practicing physicians, the higher incidence of the worst grades of prostate cancer in the finasteride group was enough to rule out the general use of the drug, despite its demonstrated promise as a preventive agent. One suspects that for most physicians the decision against finasteride was not a pencil-and-paper calculation, even if the finasteride literature is written largely in the language of numbers. To prescribe a drug that appears to increase the most ominous grades of cancer simply seems contrary to good practice. Or at least to prescribe the drug preventively—that is, speculatively—seems so.

To add to the ironies of a story ironic enough as it is, untold numbers of men have been and are taking finasteride under the trade name of Proscar at the same dosage as in the PCPT, but as a treatment for enlargement of the prostate. Doctors who will not

.ribe the drug without a clear and compelling medical reason
prescribe it for BPH. Most of them probably prescribed it before the risks of finasteride became well known with the results of
the PCPT. Only weeks before the publication of those results, an
article appeared in *Urology* declaring the use of 5 alpha-reductase
inhibitors in the treatment of BPH "safe" but recommending
against their use in BPH patients without symptoms. The author
concludes that "the concept of, 'first, no harm to our patients' has
never been more evident than in unnecessarily and imprudently
treating asymptomatic men with BPH." How much more would
such reasoning apply to men without BPH and without prostate
cancer taking the same drug at the same dosage, as a preventive,
now with its safety in question?

Some thirty-five years ago the synthetic hormone DES (diethylstilbestrol) was banned as a pregnancy drug by the FDA after being prescribed for decades in the mistaken belief that it
prevented miscarriages and somehow made for a healthy pregnancy. DES, it turns out, was a dangerous carcinogen that posed
especially dire risks to the daughters of the women who took it. In
1953 its efficacy had been found wanting in a controlled (but
poorly designed and statistically underpowered) clinical trial, a
finding supported by five other controlled trials between 1950 and
1955. Yet many obstetricians went right on recommending it. The
reason for their disregard of the data would seem to be that they
were so accustomed to using the drug that they were just not impressed by the counterevidence of the clinical trials. At the time,
"DES had already become a standard of care for certain at-risk
pregnancies and was deeply entrenched in obstetrics practice."

Consider the similarities and dissimilarities to finasteride.
Unlike DES, Proscar does what it is supposed to do—shrink the
enlarged prostate and relieve urinary symptoms—which is surely
why it was so enthusiastically received. By the time the PCPT
opened for enrollment in 1993, Proscar was already being used by
a quarter of a million men in the United States, and by the time

the equivocal conclusions of the study came to light a decade later, the drug was that much more deeply embedded in American urological practice. Where DES remained in obstetric use after the 1953 trial (in the midst of the baby boom) but probably would never have gone into use if the trial had come first, Proscar remains in use after a trial, this one well designed, that discovered that it may induce cancer: a trial that might well have put a stop to the drug if it had not already been on the market. (It does seem, however, that fewer urologists are prescribing Proscar now than before.) Possibly the urological profession reads the PCPT with a bias something like that of obstetricians in the 1950s who ignored data regarding a drug they were in the habit of prescribing. The message that finasteride appears to promote high-grade cancer cannot be a welcome one to those who have been prescribing it to multitudes of men in the belief that it was quite safe. The saving difference is that the urologists, whether or not instructed by historical experience, have corrected against their own investment in finasteride to the extent of refusing to recommend it for prevention.

The DES story is all the more troubling in that the drug's dangers were suspected long before they appeared. In 1939, shortly after it was synthesized, the Council on Pharmacy and Chemistry of the American Medical Association cautioned in the *Journal of the American Medical Association* that DES "may be carcinogenic under certain conditions," concluding that "its use by the general medical profession should not be undertaken until further studies have led to a better understanding of such drugs." Soon enough, however, DES was in general use. The DES story warns, then, not only against an overinvestment in existing practice but against the willful disregard of available evidence. And as the evidence now stands, the conclusion of the Council on Pharmacy and Chemistry applies almost word for word to finasteride.

It took about 15 years after the DES clinical trial in 1953 for the drug's risks to reveal themselves, and not until 1971 was its use by pregnant women banned by the FDA. A world in a hurry needs to remember that, as two researchers have written, "clinical trials usually concentrate on the therapeutic effect of a treatment, which tends to be a shorter term effect than carcinogenic effects that may take years to manifest." The carcinogenic effects of finasteride, if any, thus may not make themselves known in full for some time; after all, the long latency period of prostate cancer makes it a natural target for chemoprevention in the first place. And those effects would of course strike more and more men as the drug was used more and more widely—a simple principle that had profound implications in the case of another Merck drug, the now-notorious arthritis medication Vioxx. Although in a randomized clinical trial the absolute difference between the number of "adverse cardiovascular events" suffered in the Vioxx group and a group given naproxen (a nonprescription anti-inflammatory) was only 1.5 percent, in the five years Vioxx was on the market it is estimated that it was responsible for tens of thousands of heart attacks. As the table above records, in the PCPT the absolute difference between the number of high-grade tumors in the finasteride group and the control group was 1.3 percent. While that doesn't sound like much, the magnification effect that set in with the use of Vioxx by millions would also come into play if finasteride went into general use, as some models assume, among the 28 million men over 50—men who would take the drug, moreover, for years on end. "Successful [chemopreventive] agents must . . . be very safe because of the very long duration of their application in men at risk of prostate cancer." Some in the field of chemoprevention have begun to chip away at this standard, though for now it remains in effect.

Even though the figure of a 25 percent reduction in the incidence of prostate cancer seems to argue in favor of finasteride, practicing urologists were not persuaded, perhaps because they know the human costs of advanced prostate cancer too well. The fallacy of confusing numbers with persons comes out with blazing clarity in an Orwellian statement in a study of the dissemination of the PCPT data. Describing a bar graph of Prostate Cancers Detected in the PCPT publicity package, the authors report that it shows that "men [in the PCPT] diagnosed with prostate cancer prior to the end-of-study biopsy saw a benefit from finasteride." Surely the authors don't mean what they said. The men diagnosed with cancer did *not* enjoy the preventive benefits of finasteride, and if their cancer was high-grade, may in fact have suffered the drug's adverse effects. The authors must have meant that the finasteride group as a whole yielded fewer diagnoses of cancer in for-cause biopsies (435 versus 571 in the placebo group). The question of the relation of persons to numbers haunts the entire finasteride debate.

If physicians did not decide the finasteride question by comparing sets of numbers, that is nevertheless exactly what the authors of the *New England Journal of Medicine* paper urged them to do. Physicians, they wrote, can use their data

> to counsel men regarding the use of finasteride. It is important to stress that finasteride reduced the risk of prostate cancer in a clinical trial marked by frequent monitoring for disease and was associated with an increased risk of diagnosis of high-grade prostate cancer. For a man considering using this medication, the greater absolute reduction in the risk of prostate cancer must be weighed against the smaller absolute increase in the risk of high-grade disease.

In this revealing passage, professional skepticism and habits of caution contend against an ardent investment in finasteride. After conceding the elevated risk of high-grade cancer (but scaling it down to a risk of "diagnosis"), the authors not only recommend but actually command all concerned to weigh a larger quantity against a smaller one, as if such an exercise could settle the finasteride question—as if finasteride lowered the cancer risk overall. They write virtually as advocates of finasteride, taking the sting out of their study's most disturbing finding by discounting the elevated risk of aggressive cancer as an artifact of the study itself (specifically, its artificially intensive monitoring regimen) and by emphasizing that the gains of finasteride exceed the risks. The model of weighing risks against benefits in the balance-scale of medical judgment not only excuses harm too lightly but cannot really accommodate a case as uncertainty-filled as the finasteride question. But in the cited passage, "weighed" also seems to mean "thoughtfully evaluated." By saying that the data must be weighed, the authors perhaps concede that an element of reflective judgment enters into the question after all, that whether or not to use finasteride is not simply a matter of computation, that we can't in this case subtract one quantity ("absolute increase") from another ("absolute reduction") because the two aren't comparable. If you can't compare apples and oranges, still less can you compare apples and hand grenades.

The following is one way to look at the finasteride question:

ESTIMATED BENEFIT AND RISK FROM FINASTERIDE
ON DEVELOPMENT OF PROSTATE CANCER

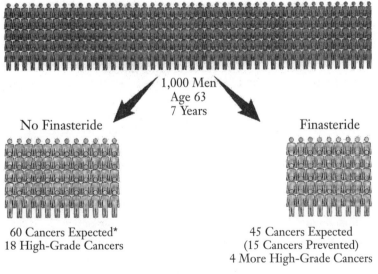

No Finasteride

1,000 Men
Age 63
7 Years

Finasteride

60 Cancers Expected*
18 High-Grade Cancers

45 Cancers Expected
(15 Cancers Prevented)
4 More High-Grade Cancers

*Probability estimate from NCI Surveillance program 1998–2000

Performing straightforward calculations on the data, some of the authors of the original report on the PCPT later arrived at the somewhat different finding that "for every 4.6 disease occurrences that are prevented [by finasteride], there is only one additional occurrence of high-grade disease"—as if this spoke in the drug's favor. But surely this sort of marginal preponderance of beneficiaries over casualties is not what the world means by prevention; and surely the word "only" is gratuitous, implying as it does that one case of high-grade cancer pales into insignificance next to so great a number as 4.6. If medicine were not held to a higher standard than many other fields of endeavor, several gains for every loss would look very good. But it is held to a higher standard, and rightly so. A recent paper in a cancer journal took as its assumption that "if chemoprevention for ovarian cancer were provided to all women over the age of 50, side effects would have to be minimal in order to achieve an acceptable ratio of benefit to risk." Those looking into the chemoprevention of prostate cancer seem to envision men starting on the drug around the age of 50 (the baseline for some economic analyses), but partisans of the drug have somehow diluted the meaning of "minimal" to the point where one aggressive cancer caused for every few cancers prevented has become, for them, an acceptable ratio. Drugs pose risks, but so does the numbers game itself.

In the twentieth century wonder drugs came to be known as magic bullets, a metaphor minted by Paul Ehrlich (1854–1915), inventor of the first promising synthetic drug, Salvarsan. Guiding Ehrlich in his research was a vision of (in his own words) "chemical compounds which have a strong destructive effect upon . . . parasites, but which do not at all, or only to a minimum extent, attack or damage the organs of the body." Ehrlich was not looking for a drug that would help more patients than it harmed, but for one that did not harm. As it happens, Salvarsan and its variant Neosalvarsan did not meet this test. But today, even as the hope of anti-cancer drugs that would work like magic bullets continues to play in our minds despite these agents' failure to materialize, we retain something like Ehrlich's standard. Even the authors who arrived at the figure of 4.6 know that a drug that spared 5, or for that matter 15, cancers for every aggressive cancer it induced could not be endorsed, all the more if the spared cancers were of doubtful clinical significance to begin with. They know that a convincing case in favor of a preventive drug would not have to explain away disturbing findings or portray aggressive cancers as incidentals. They recognize that, as a European team reviewing the PCPT data noted, "the potential mortality resulting from one high-grade tumour might equal that from several of a lower grade, and a greater incidence of more aggressive disease could outweigh any benefit from an overall reduction in cancer." And because they do know all this, they stop short of wholeheartedly endorsing finasteride despite finding that its benefits statistically outweigh the risks, whether by 4.6:1 or even 16.3:1. So it is in the original *New England Journal of Medicine* paper. The authors' leaning toward finasteride is restrained by an awareness of the drug's disqualifying risks, by a recognition that troubling data must be reported to the world accurately and in full, and by an ethos of caution—in short, by their intellectual honesty and medical principles. An ethos of caution has also restrained physicians from recommending finasteride as a preventive (when they were already, after all, using it as a treatment). The finasteride story therefore speaks of the continuance of crucial ethical habits even in a high-speed age.

3. CONSIDERATIONS OF HARM

It is because the authors respect the precept Do No Harm too deeply to gloss over harm that the original *New England Journal of Medicine* article does not come down unequivocally in favor of finasteride on the grounds that it spared more cancers than it caused, if indeed it caused them. (That the authors do not actually cite this precept only means that they respect it implicitly, that it goes without saying.) So too, it is because they care about harm that they suggest, in finasteride's defense, that the numbers in fact overstate the harms associated with the drug.

When the prostate is biopsied, small "cores" of the gland—in the PCPT, at least six—are removed for study. It stands to reason that the smaller the gland, the better the chance of locating cancer in a given number of tissue samples; conversely, the larger the gland, the greater the space cancer can hide in. Because finasteride, like dutasteride, shrinks the prostate (indeed it was this property that led to its being tested as a cancer preventive in the first place), it is possible that the finasteride group would show higher rates of cancer, and particularly of high-grade cancers, not because the drug actually caused such malignancies but because of the very nature of biopsy. As the authors write,

> We recognized the possibility that there would be an increased number of positive biopsies among men who received finasteride, because a proportionately greater volume of gland would be sampled from smaller glands. This effect could introduce a bias against any evidence of benefit from finasteride.

Unlike many of the findings of the PCPT, which could hardly have been foreseen, the possibility of a bias that would work against finasteride concerned the study's designers from the beginning. In 2000, before most of the end-of-study biopsies took place, they tried to correct for sampling bias by recommending changes in biopsy technique. The possibility that the most disquieting finding of the PCPT might actually be a mirage resulting from the nature of sampling (or even from certain mathematical adjustments introduced in the course of the PCPT) is investigated elsewhere in the literature—further testimony to the importance of considerations of harm in the finasteride debate.

In a 2006 paper in the *Journal of Urology*, Canadian investigators following up on the PCPT concluded that "the increase in higher grade tumors among men in the finasteride arm of the PCPT may simply result from prostate volume reduction." (As with the word "may," the authors tread carefully with so much at issue.) In biopsy of the prostate,

> the volume of each needle core is constant and the sextant approach [that is, the taking of six samples] will therefore proportionately sample more gland as the size of the prostate decreases. Any Gleason pattern [not to be confused with Gleason *score*] 4 or greater would have a higher chance of being sampled in small glands. The chance of sampling a section of high grade cancer in larger glands in thus reduced. . . . Sampling of the dominant (i.e. higher grade) lesion in smaller glands would be more probable than in larger prostates where the probability of sampling a nondominant cancer is higher.

Whereas biopsy samples represent only a fraction of the gland, prostate glands surgically removed are fully available for study under the microscope, like open books. That cancer rates at this stage no longer vary according to the size of the gland confirms that elevated rates of cancer in smaller glands *at biopsy* are an illusion. Only if all nine thousand final participants in the PCPT had agreed to have their prostates removed for inspection would the study's authors have been able to avoid the errors inherent in biopsy and achieve an accurate measure of cancer actually present.

The Canadian paper does not argue, but nevertheless strongly implies, that if the finasteride debate is ever resolved in the drug's favor, it will be when people generally agree not that finasteride does good despite its harms, but that it does not do harm. Certainly the paper holds out the possibility that the most dramatic harms associated with finasteride in the PCPT are exaggerations, that the numbers standing in the way of the drug's acceptance have a more innocent explanation. As the authors note,

The implications of this study [of grading bias] are even more important in light of the results of the PCPT. If one assumes that low grade cancers are effectively prevented [by finasteride] and that high grade cancers are not, one would expect, based on a biopsy derived grading artifact, a higher proportion and number of high grade cancers in the patients treated with finasteride, whose prostates were subject to a . . . 24.1% relative reduction. This, in fact, was observed.

On this showing, finasteride doesn't cause high-grade cancer but simply shrinks the prostate, making it more detectable, indeed disproportionately so, at biopsy. The authors do not comment on the irony of an expectation that no one in fact expected, or on the irony of a drug that effectively prevents the less dangerous but not the more dangerous cancers.

Also unremarked on is the ironic conclusion of another review of the PCPT:

Until biomarkers or other indicators are developed that can reliably distinguish between indolent and clinically significant prostate cancers, the true benefit of finasteride chemoprevention will be reduction in cost of diagnosis and treatment of cancers that were never likely to produce morbidity or mortality.

It is a curious drug whose greatest attribute is that it prevents clinically insignificant disease. And even this recommendation crumbles, for as the same paper duly notes, many physicians are reluctant to prescribe finasteride, a costly drug with "potential side effects," precisely because the cancer it prevents may not be worth preventing. The discussion in question leaves a confusing impression but does manage to shift the discussion from finasteride's harms to its harmlessness.

The gold standard of prevention is the vaccine, and it bears remembering that in the eighteenth century when smallpox inoculation began in Britain, "the secretary of the Royal Society of

London compiled statistics to show that the balance of probability was that inoculation saved lives," just as William Osler on the eve of World War I urged any thinking person to compare the mortality statistics of the vaccinated German army against those of the other armies of Europe. Advocates of finasteride still refer to the mortality tables—a resemblance that points to the endurance of tradition in an age that has little good to say about tradition. But the pertinence of the story of smallpox inoculation to the finasteride question lies not so much in the use of statistical arguments in each case as in the force of countervailing arguments. Although Daniel Bernoulli calculated that inoculation against smallpox would increase average life expectancy by two years, another mathematician, the *philosophe* Jean d'Alembert,

> questioned whether most people were actually prepared to run a significant risk (say one in a hundred) of immediate death in order to gain only two years. The state and society might gain if everyone was inoculated, but d'Alembert had considerable sympathy with cowardly individuals who had no desire to put their lives at risk by a deliberate act. He saw himself as taking the viewpoint of a mother of a small child rather than a father. You could not reduce, d'Alembert argued, the sort of decision involved in deciding to be inoculated to a simple cost-benefit analysis. There was no simple ratio that would allow one to weigh an increased life expectancy against a risk of immediate death.

In the case of finasteride, while the numbers prophesy an overall increase in life expectancy (though nothing approaching a full two years) if the drug went into general use, the risk of aggressive cancer is high enough not only to countervail those general benefits but to throw into question the fitness of the balance scale itself as the emblem of judgment. No simple ratio allows one to measure the benefits of preventing prostate cancer against a risk of invasive disease.

In d'Alembert's time there was another factor confounding measurement of the benefits of inoculation: inoculated persons

themselves became infectious. Smallpox inoculation thus drama-
tized what is now called iatrogenesis, the production of harm by
medical measures themselves. Many forms of iatrogenesis are still
at work, among them the use of drugs with adverse effects. But if
"the toxicity of many new agents [that is, drugs] has been discovered only after they were in regular use," how would it look if the
medical world were to endorse finasteride as a chemopreventive,
fully aware of its showing in the PCPT, only to discover that it fuels aggressive cancer just as the PCPT suggested? In that case it
would certainly appear as if a drug had gone into general use *after* its toxicity had been discovered.

Vioxx stands as a dramatic example of a drug whose dangers
appeared only when it was already in general use. (Coincidentally,
Vioxx too figures in the prostate cancer chronicle. When it was
withdrawn from the market it was beginning to be studied for
prostate cancer prevention.) A somewhat different case is DES,
the synthetic estrogen taken by some millions of pregnant women
from the late 1940s to the mid-1970s, only to see their risk of
breast cancer increase, their daughters visited with vaginal and
cervical cancer, their sons with infertility and genital defects.
While DES was suspected as a carcinogen almost from the beginning, the full magnitude of its dangers revealed itself only
some three decades after the drug was first produced. As it happens, DES was originally used in the treatment of prostate cancer;
indeed, it was the first synthetic therapy for cancer at all. One of
those who developed it, Charles B. Huggins, ultimately received
the Nobel Prize in medicine. (Again coincidentally, it was in a series of clinical trials of DES treatments for prostate cancer that
the Gleason grading system came into use.) DES poses an especially vivid warning of the magnification of a drug's dangers when
it comes to be used on a large scale—in this case, several million
women in the United States alone. While no one in the finasteride literature mentions such disasters, it is surely the prospect
of the use of a carcinogen by millions of men that deters almost
everyone from endorsing the drug despite indications that its

dangers may be more apparent than real, and despite the innocuous explanations, including detection bias—the more frequent
discovery of existing cancer in smaller glands—of the elevated
rate of aggressive cancer on the finasteride side of the PCPT.

Defenders of finasteride often cite detection bias to explain
away its association with aggressive cancers. No one can or does
argue that detection bias excludes the possibility that some of the
marked increase in high-grade cancer on the finasteride side is
real. Detection bias and real biological effects are not mutually
exclusive alternatives, though they are discussed as if they were.
There may also be subtle biases in the finasteride papers working
in the other direction—*against* acknowledgment of the drug's
dangers. Many of the authors of the finasteride papers have dis-
closed some financial indebtedness to Merck, the manufacturer of
Proscar. Of the five authors of a paper on the PCPT that appeared
in the *Journal of Clinical Oncology*, three report serving as consult-
ants to Merck, one receiving honoraria from Merck. (Other
prominent researchers have financial ties to the maker of Avo-
dart, that is, dutasteride, another and possibly more potent
5 alpha-reductase inhibitor now in clinical trial as a chemopreventive.) In addition to such interests is the authors' professional
investment in the safety of the drug in question. By the time the
results of the PCPT came out, urologists had already been prescribing finasteride at the same dosage for BPH for a decade.
Would they not prefer to consider such a drug safe? What doctor
would want to think that a drug prescribed for a *benign* condition
was itself prohibitively dangerous? But because almost all the finasteride authors are committed as a matter of principle to erring
on the side of caution, they withhold a verdict in favor of finasteride despite strongly leaning to the opinion that the drug really
is both beneficial and safe.

The claim that the finasteride numbers are distorted by a detection bias, while certainly plausible, gives pause. If the biopsy of
a smaller gland yields a higher rate of positives, the entire PCPT,

with its methodical, literally microscopic search for cancer, could be likened to a biopsy of a small corps of men representing a segment of the male population in the United States. Because the surveillance regimen of the PCPT itself was an exercise in detection bias, it discovered cancer at such an inflated rate that even the suppressing effect of a potent drug was not quite enough to bring it down to "normal." "PCPT suggests that in a well-screened group, the proportion of diagnosed tumors of no risk to the patient may be as high as seven out of eight." While the PCPT was artificially rigorous, in medical practice itself prostate cancer has been searched out so actively that it is being found in vanishingly small amounts. "With more vigilant screening for prostate cancer, there has been an associated increase in patients with little or no residual cancer at radical prostatectomy after an initial diagnosis of minute cancer on needle biopsy," suggesting that the biopsy itself has removed whatever cancer is there, and that the gland has been excised for nothing, literally. Detection has reached the vanishing point. Detection bias thus cuts two ways: if to some it points to the conclusion that the numbers of more aggressive tumors in the finasteride side of the PCPT look worse than they really are, to others it will suggest that the more persistently you search for prostate cancer the more likely you are to discover it—whether or not it would ever come to light of its own, and whether or not it is really harmful—and that such cancer is therefore better left alone. The caption of the PCPT might be: Beware what you look for—you might find it.

4. Salutary Restraint

A few years after the appearance of the original *New England Journal of Medicine* paper, a number of the authors collaborated to argue their case more fully, concluding that "increased detection due to reduced gland volume contributed to the finasteride-associated increase in high-grade disease." Readers were reminded that "the increased risk of high-grade disease on finasteride in the PCPT . . . was noted in the first year [when shrinkage of the gland occurred] and did not increase over time, raising the suspicion that the increase may have been due to other causes besides truly induced aggressive disease." (The assumption was that if finasteride were responsible for high-grade cancer, it would be exerting its effect throughout the PCPT.) Such circumstantial evidence served to counter, but not altogether defuse, the now common objection that finasteride is just too dangerous to use preventively. While the authors were prepared to argue that the original data may actually have understated finasteride's preventive effect, they conceded, on the other hand, that "the evidence does not rule out that finasteride may have induced high-grade prostate cancer in some men," a risk most will deem prohibitive. It is only because the entire finasteride debate is framed by the unvoiced precept Do No Harm that the drug's defenders found the burden of argument squarely upon them, their tone somewhat chastened as they were compelled to admit that they really do not know quite what the drug does. In some corner of their minds the original researchers may have said to themselves, "It is unfair that we, the first to advance a real hope for the prevention of prostate cancer, the ones who got things going, should be thrown on the defensive." But it is better thus. If not for the restraining effect of the rule Do No Harm, we might well be misled by our own hopes into the use of a dangerous drug in a big way. In the face of public impatience to get drugs tested, approved, and marketed—the National Cancer Institute has a program known as RAPID to hasten the synthesis and testing of chemopreventive drugs—the restraint was found not to put to general use a medication, finasteride, already approved and widely used for another purpose.

One benefit of the prevention of prostate cancer would be the suppression of many of the more questionable cancers now detected in great numbers, and, once detected, treated because of

the chance they would prove aggressive. Why do urologists recommend screening for prostate cancer when screening results in
overtreatment, with all its attendant costs and miseries? Some
would say they are covering themselves legally. The author of a
trenchant critique of medical overtesting tells of a colleague who
failed to order a PSA for an asymptomatic patient who a few
months later was found to have advanced prostate cancer, and
who then sued the urologist and won. Doctors, the author comments, "are always worried about missing something and having
someone else find it," especially when the something is cancer
and the someone is a lawyer. But the fear of being found out does
not apply to the use of finasteride, because if the drug did somehow induce invasive cancers, they could not possibly be traced. If,
among one thousand men taking finasteride, twenty-two high-
grade cancers appear where only eighteen would be expected, no
court of law on earth would be able to establish which were the
surplus cancers owing to finasteride. Doctors reluctant to prescribe finasteride broadly cannot therefore be deterred by the fear
of exposure. They must be deterred not by the thought of being
discovered to have done harm, but by the thought of doing harm.

Urologists know full well that PSA testing leads to overdiagnosis, that it is a rough measure, that its levels of PSA defined as
high are artificial, that it yields too many false positives and false
negatives, that it is just too Procrustean an instrument. Yet they
remain committed to the PSA as the only available means of detecting prostate cancer while it is still without symptoms. The
contrast with British practice is telling. Perhaps because they do
not wish to introduce the sort of epidemic of diagnoses that PSA
testing has brought to the United States, where the incidence of
prostate cancer runs some *seven times* higher than in England and
Wales, British physicians have taken a dubious view of the PSA.
So comparatively unfamiliar is the test in Britain that the National Health Service recently composed a primer to introduce

men to it, and so warily is the test regarded that this document carefully offsets every benefit with a risk. In the words of the primer, the risks of the PSA test are that

- It can miss cancer, and provide false reassurance
- It may lead to unnecessary anxiety and medical tests when no cancer is present
- It might detect slow-growing cancer that may never cause any symptoms or shortened life span
- The main treatments of prostate cancer have significant side-effects, and there is no certainty that the treatment will be successful

As if this weren't enough, the pamphlet also details the next step in detection, the biopsy, along with its risks.

The finasteride papers are written with something like the rigorous skepticism of the NHS flyer. That is, although American medicine attacks prostate cancer more aggressively than its British counterpart, the Americans assume a conservative position when the question of using a powerful but questionable and poorly understood chemopreventive drug stands before them. Like opponents of the test which for them represents the only hope of early detection, they sound notes of caution and doubt. But perhaps language misleads us, and it is the Americans' very aggressiveness toward prostate cancer that counsels caution in the use of finasteride. If, faced with the enigmas of prostate cancer, American clinicians "often follow the safest approach and recommend aggressive therapy," how much sense would it make for them to throw safety to the winds for the sake of Proscar? Clearly, in any case, it is because the risks of using a possible carcinogen on a general scale are not to be trifled with that the medical literature approaches finasteride so guardedly. Its wariness is well founded—all the more because of the possibility of other means of chemoprevention not as risky.

The possibility that selenium and/or vitamin E may contribute to the prevention of prostate cancer is now being tested in a trial involving some 32,000 men, though the rationale for the study has been cogently questioned. Another attractive candidate for the chemoprevention of prostate cancer is a drug, or family of drugs, already in use for another purpose, but apparently without the disqualifying liabilities of finasteride. Statins, now widely prescribed as cholesterol reducers, "first attracted interest for cancer prevention as an unexpected result" of clinical studies of their efficacy in preventing cardiovascular disease. Initially suspected of increasing the risk of cancer, statins, upon investigation, were found to be associated with "provocative and unexpected" declines in the incidence of certain cancers, a decrease in cancer generally, and a marked decrease in aggressive prostate cancer. These benefits, however, are contested. The best that can be said is that statins *may* reduce the very grades of cancer whose elevated numbers have ruled out the preventive use of finasteride. In a recent large study of the effects of statins and other cholesterol reducers on the risk of prostate cancer,

> There was no overall reduction in prostate cancer risk, but a provocative analysis of the extent of the disease showed a significant (46%) reduction in advanced prostate cancer risk (compared with non-drug users), and the risk decreased with increasing duration of use. . . . The risk reduction was even greater for metastatic and fatal disease. Although the study involved drugs other than statins, the strongest risk reductions occurred towards the end of the study, when 90% of the participants who were using drugs to reduce cholesterol levels were taking statins for this purpose. . . . [In addition], a recent case-control study conducted by the Veterans Affairs system has shown that statin use is significantly inversely associated with overall prostate cancer risk and is strongly inversely associated

with high-grade prostate cancer; these reductions increased with prolonged statin use.

If, as seems possible, finasteride doesn't actually promote aggressive cancer but has merely the perverse effect of inhibiting the less dangerous but not the more dangerous forms of the disease, statins may inhibit high-grade disease preferentially. How different the story would be if finasteride were such an intelligent and considerate compound.

5. The Return to Evidence

To a layman it comes as a surprise to learn that some argue for a thing called "evidence-based medicine"; for on what (one asks innocently) is medicine based if not evidence? All sides agree that evidence of the highest quality, probative evidence, is obtained from randomized clinical trials; yet twenty years after the inauguration of PSA screening, the value of PSA has not been determined in such a trial. The test itself has not been tested. One physician notes, "It is a fact that screening *may* save lives, but it also a fact that screening has not been proven to save lives and that there are some definite harms associated with screening. . . . Rarely is it emphasized that the definitive studies of prostate cancer screening have not been done." Even so, PSA continues to drive prostate cancer medicine. In part it is because of the flood of diagnoses of prostate cancer let loose by PSA testing—by no one's design—that the need for some means of preventing the disease became so acute. From PSA testing came overtreatment— too much surgery, too much radiation—and the best escape from this bind would be to keep the disease from occurring in the first place, or at least to cut its incidence. Thus the PCPT; but the PCPT in turn yielded findings few dreamed of when finasteride came into use fifteen years ago. Although clinical studies of the drug's efficacy as a BPH treatment—some lasting only days, and none apparently randomized—were conducted before it went to market, it is hard to conceive that finasteride would ever have been adopted as a treatment if it had been determined at the time to be a possible carcinogen. (When the author of these preliminary studies learned in 2003 of the increase in high-grade cancer attributed to finasteride, she told the *Wall Street Journal* that the issue "clearly requires further study.") In these respects prostate cancer medicine has drifted out of compliance with the model of medical practice governed by the best standards of evidence. Those who refrain from recommending finasteride for prevention, in response to the PCPT data, are returning in effect to evidence-based medicine.

The findings of the PCPT, both pro and con, are so arresting that one can scarcely *not* pay attention to them. The many who have sought to resolve or dissolve the paradoxes of the PCPT, contending that finasteride's carcinogenic effects are more apparent than real, speak directly to the data. While their arguments are both forceful and ingenious, sometimes the paradoxes of cancer are real. Prostate cancer may be both androgen- and estrogen-driven. Certain compounds of interest in the chemoprevention of prostate cancer have both estrogen and anti-estrogen effects, as does tamoxifen. "Some agents are preventive in one organ and carcinogenic in another (*e.g.*, tamoxifen is preventive in the breast, carcinogenic in the uterus). . . . Some agents can be preventive and carcinogenic in the same organ." It may be that finasteride's carcinogenic effects as registered in the PCPT are in part illusory and in part real, and that it too is both preventive and carcinogenic, just as it seems to be—which is not to say that its effects somehow balance out. Unless and until the shadow of doubt cast over the drug by the PCPT is lifted, the evidence argues that it must be considered too dangerous to use in a preventive, that is, speculative, manner.

And this is exactly how urologists do consider it, guided as they are by the principle of avoiding harm. Indeed, *only* that principle stands in the way of their prescribing this drug with so much in its favor except that it appears to fuel aggressive cancer. Finasteride cuts the incidence of prostate cancer significantly— possibly by even more than its 25 percent showing in the PCPT; it correspondingly reduces overtreatment of the disease, with its costs and sorry side effects; and it is the only drug that can do all this. Those who gain by it outnumber the losers. And these impressive, some would say compelling, arguments in favor of finasteride are overthrown by a single counterargument: the higher rate of dangerous cancer associated with the drug. That consideration alone has kept the drug on the shelf. Although the traditional maxim Do No Harm has been questioned, ridiculed, and quietly or not so quietly set aside, in this case it continues to exert its deterrent force. Over the course of these pages I hope to bring out the extent of medicine's dependence on tradition—an undervalued source of guidance and cultural wealth—in its approach to the finasteride question.

[3]

Sister Drugs:
Finasteride and Tamoxifen

❦❦❦

1. WEIGHTS AND MEASURES

Prostate cancer and breast cancer are parallel universes—strangers to each other but twins. While for a patient each disease is a world unto itself, and while a man with prostate cancer is unlikely to seek the advice of a woman with breast cancer, or vice versa, in fact these conditions are linked by all kinds of family resemblances. Indeed, breast cancer genes are suspected to be involved in prostate cancer.

As I wrote a few years ago in *Seeds of Mortality*:

In the case of both breast cancer and prostate cancer, the wisdom of aggressive screening has been questioned, meaning that those confronting these cancers, either in fact or in prospect, are up against a disease more subtle than our instruments of detection and in some ways still beyond the reach and understanding of medicine. . . . In both cancers, it seems, evidence of the disease shows up in autopsies of those who died of other causes. In Japan, rates for both diseases run lower than in the United States. Two or three years before my diagnosis [with prostate cancer], a neighbor two or three years older was struck with breast cancer. Both cancers are "among the most malignant and clinically intransigent" forms of the disease. Like mirror conditions, prostate and breast cancer are diagnosed in about the same numbers and account for about the same number of deaths per year in this country.

Additionally, in both cases testing leads only too often to a finding of cancer, treatment with its attendant effects, and a treadmill of anxiety, expense, and more testing. In both cases many, perhaps most, discovered malignancies would not turn deadly if left to themselves, but no one can say of any given malignancy that this is the case. "Somewhere between 50 and 80 percent of DCIS [a commonly detected form of incipient breast cancer] and screen-detected prostate cancers . . . are pseudodisease—but we can't say just which ones," leaving the patient, as ever, playing Russian roulette. In both cases, though a family history of the disease puts one at higher risk, most patients have no such history. Breast cancer patients stare at the disfigurement of a radical mastectomy (now performed less often than before, however), prostate cancer patients at the wretchedness of a radical prostatectomy. Considering all this kinship, it seems fitting that the PCPT should have been preceded by the BCPT—the Breast Cancer Prevention Trial. Indeed, the latter may have inspired the former. "The promising results of the . . . breast cancer prevention trial . . . clearly indicate that breast cancer prevention is possible. This suggests that prostate cancer prevention is also possible."

In the manner of yin and yang, the PCPT tested a drug that targets testosterone, the BCPT a drug—tamoxifen—that blocks estrogen effects. Both agents were in use for another purpose while they were tested as preventives, and both remain poorly understood in spite of being both used and studied. (Certain prostate cancer patients have themselves been treated experimentally with tamoxifen.) Both proved to reduce the general incidence of cancer but not more aggressive cancer—"high-grade tumors in the case of finasteride; estrogen receptor-negative disease in the case of tamoxifen." Both are viewed warily and recommended for high-risk patients if at all, despite the prevalence of the respective cancers in normal-risk populations. Like the PCPT, the BCPT was halted a year early, having already mathematically demonstrated what the experiment's designers hoped to establish—in this case, the preventive effect of tamoxifen. Also like the PCPT, the findings of the BCPT were disturbingly equivocal all the same, giving rise to both joy and troubled contemplation. In one and the same *New York Times* story on the BCPT, "jubilant" persons acclaim the results of the clinical trial as "historic" while others reserve judgment and point out that the BCPT has "no simple take-home message" at all. It is as if while we experiment with tamoxifen, tamoxifen experiments with us to discover our contradictions. In any case, as later with finasteride, the tamoxifen findings were too ambiguous to incite a revolution in medical practice. "Despite the established preventive effects of tamoxifen and finasteride, the important risks / side effects associated with these agents produced a widespread resistance to their use for preventing breast or prostate cancer." When the celebration of finasteride as a result of the findings of the PCPT did not occur, it was the second such rain-out of a victory parade in five years. The BCPT ended in 1998.

The BCPT found that tamoxifen reduced the incidence of both noninvasive and invasive breast cancer by almost 50 percent in women at moderately elevated risk. If the 25 percent risk reduction attributed to finasteride in the PCPT was impressive, all the more was the showing of tamoxifen. (Perhaps the finasteride figure would have been higher if the PCPT had enrolled men at higher risk of prostate cancer.) But by the same token, if the prospect of reducing breast cancers by half was not enough to win the medical community over to the preventive use of tamoxifen, still less was the 25 percent figure, in and of itself, capable of bringing in a verdict in favor of finasteride. Those who make a numerical argument for the use of finasteride on the basis of the PCPT's findings fail at times to consider that numbers twice as dramatic were not convincing in the case of tamoxifen. The story of the BCPT and the PCPT alike suggests that impressive numbers are not all, and that the decision whether to use a risky drug to reduce risk is not, in the end, a matter of computation but of judgment. Tamoxifen therapy carries a risk of dangerous blood clots some thirty times higher than oral contraceptives, and yet the latter risk is considered "significant" because the women are healthy. If the risk for tamoxifen users is not labeled "extremely significant," or with some other black flag, it is because judgment is speaking.

The risk of finasteride, as we know, is an increased rate of high-grade tumors of the prostate. Suggestively, in testimony before a House subcommittee looking into the BCPT in 1992, a researcher argued that "treatment with tamoxifen stimulates the growth of a class of aggressive breast cancer tumors." This, however, seems unconfirmed. The confirmed risks of tamoxifen—disquieting enough to deter the general use of the drug—are not absolutely analogous to the risk of finasteride. The risks of tamoxifen are elevated rates of stroke, pulmonary embolism, and deep-vein thrombosis as well as a marked increase in endometrial cancer (that is, cancer of the lining of the uterus). ("Beware This Breakthrough!" ran the suitably ironic headline of a *Time* report on the BCPT in 1998.) According to the BCPT team, endometrial cancers attributable to tamoxifen are nonaggressive and manageable, though one of the principal investigators has stated that "of 23 reported uterine cancers associated with tamoxifen, only ten had a good prognosis," and a review of the data by the American Society of Clinical Oncology in 2002 found that, in one study, "long-term tamoxifen users who developed endometrial cancer had a relatively unfavorable prognosis related to histology . . . and higher stage." In contrast to the PCPT where the number of more aggressive malignancies did not continue to rise throughout the study (which some interpreters read as evidence that finasteride probably didn't in fact contribute to these cancers), endometrial cancers rose from one end of the BCPT to the other. Even so, the correspondence between tamoxifen and finasteride remains: two drugs, each promising as a chemopreventive, so promising in fact that each clinical trial was concluded early, but both clouded with deterrent considerations. The two drugs are medical cognates, as prostate and breast cancer are themselves cognates. Only because of purely journalistic reasons that have nothing to do with medicine, biology, or indeed statistics has the press paid more attention to tamoxifen than finasteride. Perhaps it was because of the press's hunger that in the case of the BCPT "less than two weeks elapsed between the decision to close the trial and public announcement, compared with three months for the PCPT"—a disparity singled out as a point of distinction between the otherwise analogous clinical trials.

Breast cancer typically strikes earlier and moves faster than prostate cancer, which may explain, in part, the higher intensity of its rhetoric. (The BCPT was open to women as young as thirty-five, the PCPT to men over fifty-five. Recall that the National Cancer Institute's estimate of finasteride's risks and benefits as a chemopreventive assumes a cohort of one thousand men *aged sixty-three*.) The issue of women's health in any case carries a political charge and a rhetoric of accusation that men's health does not, despite the almost identical lifetime risk of dying from breast and prostate cancer. Correspondingly, where the tamoxifen trial was vigorously protested before the fact and much reported afterward, the finasteride trial was neither protested nor reported with any degree of follow-up. That the cognate drugs are nevertheless handled with the same wariness by practicing physicians—being used in strictly limited ways despite their dramatic chemopreventive promise—suggests that, all in all, these doctors' fidelity to the principle of caution runs deeper than their sensitivity to journalistic and political influences.

Because of the projected number of cancers prevented on the one hand and serious, even dangerous results on the other, the

BCPT, even before it began, posed a dilemma comparable to the finasteride question. "It is a matter of opinion whether the prevention of breast cancer in 62 women [among 8,000], many of whom would have been completely curable by standard treatment, was worth putting healthy women at risk of unpleasant or even potentially life-threatening side effects." If a researcher contemplating this question in a vacuum finds it a coin toss, a researcher situated in a medical tradition whose first principle is Do No Harm is not in a vacuum. Do No Harm creates a presumption against an experiment from which significant harm is expected to flow. It is hard to imagine that the Prostate Cancer Prevention Trial would have been conducted in the expectation that it would significantly elevate the rate of aggressive cancers. The rise in high-grade tumors was not an expectation built into the PCPT but a finding alarming enough to have ruled out the preventive use of finasteride, and one whose import is still being deciphered. The suspension of the principle of avoiding harm in the case of the BCPT probably came about not because of medicine's supposed animus against women—even as the subjection of women at the hands of medicine became an established doctrine in academic circles, older men by the thousand, well outside the spotlight of concern, were being reduced to impotence by having their prostate removed—but because of the ardor and pressure of hopes invested in the possibility of preventing breast cancer.

With tamoxifen proving unsuited for general use, attention turned to identifying those patients to whom it might reasonably be recommended for prevention: women already at elevated risk of breast cancer. As with finasteride, the laws of irony dictated that the best candidates for preventive tamoxifen would be at a risk high enough to justify the risk of the drug itself. As the BCPT team put it,

> The decision relative to which . . . women should or should not receive tamoxifen for breast cancer prevention is complex. The primary determinant for making such a decision relates to each

woman's projected risk for breast cancer. Women whose breast cancer risk is sufficiently high to offset the potential detrimental effects of tamoxifen would be candidates for the drug.

Ironically too, though the BCPT was halted a year early in order to allow the placebo group the chance to take tamoxifen, many members of that group may not have been good candidates in view of tamoxifen's now-demonstrated risks. Another study, generalizing from the combined experience of the BCPT and PCPT, cautions that careful balancing of risks and benefits is likely to be necessary with any effective cancer-preventive drug:

> It is critically necessary to identify populations whose high risk of a serious cancer demonstrably outweighs the potential risks and side effects, which now seem virtually inescapable, of taking an effective chemopreventive agent.

But can the weighing of risks against benefits be reduced to a calculation?

For the sake of discussion, the BCPT team proposed one such method. "One way in which the benefit from tamoxifen can be estimated is to subtract the number of unfavorable events from the overall number of cancers prevented." But what good is such a crude calculation? "It seems inappropriate," argued the team, "to view an endometrial cancer as being 'equivalent' to a breast cancer, since, when endometrial cancers occur in women who receive tamoxifen, they are most often curable by hysterectomy and the mortality rate is minimal." A woman might not take so light a view of the loss of her womb. But if the endometrial cancers were more serious (as it seems they may well be), would that make the computation of risks and benefits any better? In that case, doctor and patient would be in the position of those trying to calculate the value of finasteride by subtracting the number of high-grade cancers induced per thousand from the number of cancers prevented per thousand. That absurd model is in fact discussed in the medical literature, along with variants giving more numerical weight to the high-grade cancers to reflect their severity. But all such calculations seem at once minutely rational and largely surreal, as if they belonged to some imaginary commonwealth of numbers descended from Plato's Republic.

An advocate of the preventive use of tamoxifen by all 29 million American women who qualified for the BCPT (almost the same as the number of men who "qualify" for finasteride, that is, men over 50) assigned a numerical value of one to all adverse outcomes, such as endometrial cancer, while conceding that "using the number of adverse events to cancel out an equal number of tumors prevented is questionable." But it is at least as questionable to cancel out actual harms with "tumors prevented." The fact is that the determination of which women should and should not take tamoxifen preventively, being (as the BCPT team acknowledged) "complex," cannot be reduced to a computation. Every attempt to do so advertises its own absurdity. Certainly a patient thinking of taking tamoxifen or finasteride preventively will consider the statistical risks and benefits, but even so, the question in each case is fraught with imponderables. Even as it approved the use of tamoxifen to reduce the risk of breast cancer, an FDA advisory committee conceded in 1998 that "it did not have sufficient information to determine which women have a risk high enough to outweigh tamoxifen's potential hazards." Moreover, when a conscientious physician contemplates a harm and a benefit, these are not strictly equal quantities even if each has a value of one according to some paper formula. The harm is weighted by the governing maxim of medicine, Do No Harm. If 29 million American women are not at this moment on a preventive regimen of tamoxifen despite the drug's proven ability to reduce the incidence of breast cancer, this is in part because physicians take the harms associated with tamoxifen, from endometrial cancer to strokes, too seriously to recommend the drug so broadly. Physicians may not speak in traditional language of their duty not to cause harm, but it is their recognition of this duty that has enabled them to use tamoxifen judiciously, which is also to say, not on a grand scale.

Ironically, the early cessation of the BCPT itself testifies to a wish to avoid doing harm—in this case by the denial of tamoxifen to women at risk. Concern for those on the placebo side of a randomized clinical trial, deprived of a potentially valuable drug, is a moral principle built into the history and foundation of the randomized clinical trial. The first such trial with human subjects, begun in 1946, tested the effect of streptomycin in treating pulmonary tuberculosis. Only because the drug was in short supply was it deemed justifiable "to carry out a trial in which one group received streptomycin whereas a control group was treated with traditional methods." Some trace of this sort of concern for the control group persisted into the BCPT, halted as it was at the point where the trial's monitors believed they could no longer justify withholding tamoxifen from those in the control group.

> It was concluded that additional follow-up would not have resulted in improved estimates of treatment effects that would have justified withholding from participants on placebo the knowledge that tamoxifen was an effective prophylactic agent. This allows those women on placebo to consider taking tamoxifen.

(The monitors of the PCPT may have come the same conclusion, though the report on that trial states merely that on February 21, 2003, "the data and safety committee met and . . . recommended early termination of the study, since the study objective had been met and the conclusions were extremely unlikely to change with additional diagnoses of prostate cancer and end-of-study biopsy results.") That monitors felt obligated to offer tamoxifen to the placebo group, even if tamoxifen is no wonder drug, seems a residue of the same concern over doing harm by withholding benefits that informed the original randomized clinical trial. In their different ways, both defenders and critics of the BCPT spoke the language of harm.

2. A Quark in Our Calculations

There is another harm to be weighed by a woman considering ta-moxifen, one lost in the discussion of the "adverse events" linked to the drug, because it lacks drama and may not even qualify as an event: impairment of one's sexuality. As the sexual side effects of finasteride would be most unwelcome to those who might stand most to gain from its preventive potential (that is, men younger than those in the PCPT, with prostate cancer in a more incipient stage), so the sexual side effects of tamoxifen "may be more prob-lematic in younger women," the same younger (that is, pre-menopausal) women recommended by risk/benefit models as candidates for tamoxifen because they are less liable to strictly medical misfortunes like uterine cancer. Even down to their co-nundrums, finasteride and tamoxifen are brother and sister.

The sexual ill effects of tamoxifen are of small concern to the medical literature because, in the eyes of the literature, they are inconveniences rather than dangers. They concern the patient more than the doctor. But there is another reason the literature has little to say about them. Precisely because they are the pa-tient's concern—precisely because only the patient can assign their weight, and for some that weight will be trivial and for oth-ers not so trivial—sexual harms play havoc with calculation itself. How to work a wild card into a computation? Granted that tak-ing tamoxifen is not as drastic as having a breast removed pre-emptively, as some genetically at-risk women do, could we blame a woman who was not so ideally rational as to jeopardize her sex-uality for the sake of reducing her risk? So personal is the evalu-ation of sexual harms that it embarrasses the very logic of computation, mocking our attempts to reduce a medical decision to a quasi-numerical procedure.

[4]

Specific Harms and
General Benefits

1. BODIES AND NUMBERS

Perhaps only in the alternate universe of prostate cancer could we encounter anything as surreal as the finding that "on average . . . men diagnosed with low-grade prostate cancer are more likely to be alive at 10 years compared with men without prostate cancer." Indeed, in Britain the average man with prostate cancer lives two and a half years longer than men without it, which makes the disease seem like a public-health benefit of the first magnitude. In the United States great numbers have prostate cancer, but small numbers die of it. Tumors range from the inconsequential to the deadly, with clinical distinctions difficult to draw and all warning signs, tests, and numerical measures unreliable in one degree or another. "In the PSA era it is possible for a perfectly healthy young man to be diagnosed with cancer, go through a diagnostic work-up, endure the adverse effects of therapy [that is, incontinence and impotence], and suffer a biochemical relapse without ever having experienced a disease-related symptom." The PSA test itself is beset with both false positives and false negatives. The meticulously designed and monitored PCPT, for its part, yielded a set of results so bewildering that they seem to mirror the finding in 2000 that the available information on the role of steroid hormones in the genesis of prostate cancer is "often contradictory."

Ironically, too, the only drug shown to prevent prostate cancer may also serve to make it more detectable, with the perverse apparent result of more disease, not less, and as if this were not enough, may suppress less harmful but not more harmful malignancies, if it does not actually induce the latter. Finasteride's most vigorous partisans hesitate to claim that it will actually save lives, and the same doctors who now prescribe it to treat a benign condition will not yet prescribe it in the same dosage as a cancer preventive. Much as, in the parlance of the medical papers, men are at risk of diagnosis, so too does the project of chemically mitigating the risk of cancer pose risks in its own right. It is in character with prostate cancer medicine that many researchers think chemoprevention should focus on groups at a sufficiently high risk to justify the dangers of chemoprevention. (Others calculate the gains in person-years that would follow if the drug were administered generally, "at the population level.") Such ironies are apt to remain with us as long as "promising interventions with biologically active substances [in chemoprevention] are likely to have adverse, perhaps unforeseen effects."

Assuming the risk of high-grade cancer to be real, if finasteride were somehow administered to a sector of the male population in the hope of reducing the general rate of prostate cancer, we would have returned in effect to the medical research of the 1950s, where "commitment to the health of the nation and, indeed, of humanity supplied the implicit justification for many medical harms to individuals." (It was as a result of such research that controls on the use of human subjects were put in place.) In the finasteride case, the distinction between specific harms and general benefits would be inflamed by the severity of the harms and the merely inferential nature of the benefits.

Those in favor of the preventive use of finasteride find themselves on the defensive not only because of the dangers of high-grade tumors but because such tumors dramatically exist, while cancers that are prevented, that do not materialize, dwell only in the phantom statistical category of "what might otherwise have occurred," and might well not have occurred even without the help of finasteride. Whether or not Smith's Gleason-eight cancer is itself due to finasteride, it is real and pertains to a person in a way that prevented cancers do not (which some will say only proves our hopeless bias in favor of persons). We can speak meaningfully of Smith's cancer but not the cancer Robinson did *not* have. Vivid harm overshadows inferred benefit. To the advocates of finasteride this too may seem unfair, but bear in mind that high-grade cancer "is often very aggressive and fast-growing, and it can be deadly." Physicians themselves know all too well what these words, written on the human body, signify. It is right that we attach special weight to such harm.

In his treatise on human nature, David Hume invests considerable importance in the difference between ideas more and less immediate and lively. At this point, as we consider the different vividness and rhetorical weight of bodily harms and statistical benefits, it may be apt to consult Hume's distinction between "two kinds of objects, the contiguous [that is, the near] and the remote." The latter, he observes,

> appear in a weaker and more imperfect light. This is their effect on the imagination. If my reasoning be just, they must have a proportionable effect on the will and passions. Contiguous objects must have an influence much superior to the distant and remote. Accordingly we find in common life, that men are principally concern'd about those objects, which are not much remov'd either in space or time, enjoying the present, and leaving what is afar off to the care of chance and fortune. Talk to a man of his condition thirty years hence, and he will not regard

you. Speak of what is to happen tomorrow, and he will lend you attention.

I propose a rough analogy. As, according to Hume's model, remote considerations exert less influence than present concerns, so statistical deductions of "cancers prevented" carry less rhetorical power than present harms. As long as elevated numbers of high-grade malignancies are attributed to finasteride (as they must be until the evidence shows otherwise), the drug's defenders will be on the wrong side of the maxim that theoretical matters appear to human imagination "in a weaker and more imperfect light" than more immediate ones. This prejudice—if prejudice it is—in favor of the present and the perceptible bedevils preventive medicine generally. According to one epidemiologist, "People are generally motivated only by the prospect of a benefit which is visible, early, and likely," unlike the benefits of a preventive regimen. "In preventive medicine the prospect of personal benefits to health provides only a weak motivation to accept a change, since it is neither immediate nor substantial, and an individual's health next year is likely to be much the same," regardless of whether that person embarks on a course of prevention like daily finasteride. It is probably because the benefits of randomized clinical trials are so remote and speculative that it can be hard to attract participants in the first place.

To some, our leaning in favor of things closest to our field of concern—"those objects, which are not much remov'd either in space or time"—only demonstrates a lamentable myopia. Commenting on the reluctance of mothers to inoculate children against smallpox, d'Alembert observed that "to enjoy the present and not to worry much about the future, that is the common mode of thought, a mode half good, half bad, but which there is no hope that men will reform in themselves." Perhaps because people were slow to reform themselves, nineteenth-century campaigners for "improvement" set out to reform them, holding up in particular the habit of taking long views, of looking to the future, as a mark of civilized behavior. Yet I would not brand someone a fool who declined to take finasteride for twenty or thirty years, with its attendant adversities, in the hope it would spare him prostate cancer at age seventy. Nor indeed would I brand one a fool who didn't get screened for the disease at all (like some of my own acquaintances), instead leaving himself "to the care of chance and fortune." For the fact is that cancer screening, and PSA screening in particular,

> can diagnose some cancers that would never have caused medical problems. A significant harm can be the unnecessary treatment of these otherwise clinically insignificant cancers. Given the toxicity and treatment-related morbidity of most anticancer interventions, detection and pseudo-cure of such lesions can only be considered pure harm—to say nothing of the harm of labeling. The mental anguish and concern caused by cancer screening should not be underestimated.

And if such a case can be made against PSA testing, how much stronger is the case against finasteride.

Although Hume's observation of the relative power of influences is not offered to the honor and credit of human nature, in some ways it is fortunate that we continue to estimate things as people did "in common life" a quarter of a millennium ago. If we approached a question like whether or not to recommend, prescribe, and take finasteride as if we were the first inhabitants of the earth to be making this sort of decision, our judgment might well lack the balance that comes of having ground—the ground of tradition—under our feet. Hume's maxim also helps explain why doctors who will prescribe finasteride to reduce an enlarged prostate will not prescribe it at the same dosage to reduce the possibility of prostate cancer. It is not that the doctors are foolish and inconsistent but that they estimate the cases differently, roughly according to Hume's maxim. The one use of finasteride is therapeutic, the other speculative. An enlarged gland is a definite thing while the statistical prospect of cancer is indefinite, like a distant view (with haze to boot)—too indefinite to justify the risks of finasteride. Additionally, untreated BPH poses risks of its own, which may justify the use of a drug whose dangers emerged only after it was approved to treat BPH. (Would doctors, knowing what they now know of finasteride, continue to prescribe it for enlargement of the prostate if that condition were simply uncomfortable and not potentially more serious? As it is, it seems fewer doctors are prescribing Proscar for BPH, while some who continue to do so regard its anti-cancer potential as a side benefit, even though they will not go so far as to recommend the drug for prevention.) In using finasteride in one case but not the other, the doctors stand in a tradition of empiricism. Their membership in tradition enables them to judge better of complex moral questions than if they rejected the past itself as one long Dark Ages, as we are often exhorted to do.

"Talk to a man of his condition thirty years hence, and he will not regard you," observes Hume, as if this proved the narrowness of human interests. But we little know what our condition will be thirty years hence. A physician who imagined he could tell the future, who was without respect for the unknown, would lack the humility before the unknown that is essential to the practice of medicine. In the case of finasteride, the anticipated benefit of prevention might or might not materialize while the beneficial effect on an enlarged prostate is proven and pronounced. Arguably, the physician who for now prefers the immediate to the speculative benefits of the drug, even though the speculative benefit is the suppression of cancer, shows cautious judgment similar in spirit to traditional prudence. We deliberate, says Aristotle in his *Ethics*, about

> the things that are brought about by our own efforts, but not always in the same way . . . e. g., questions of medical treatment. . . . Deliberation is concerned with things that happen in a certain way for the most part, but in which the event is obscure, and with things in which it is indeterminate. We call in others to aid

us in deliberation on important questions, distrusting ourselves as not being equal to deciding.

The medical literature on finasteride is a forest of indeterminacy, through which physicians who prescribe the drug for BPH but not for the prevention of prostate cancer are tracing a path.

Unlike malignancies prevented by finasteride, with their purely inferential existence, the benefits of finasteride as a BPH treatment accrue to definite persons (as do high-grade malignancies, for that matter). The physicians more impressed by effects registered in actual persons than by abstract statistical benefits are still connected to the venerable tradition that thinks *of* persons, not just of categories, abstractions, or statistics: a tradition that includes virtually the whole of literature. In the face of powerful tendencies in our world, and in medicine, to dissolve persons into units or members of categories, the persistence of the tradition of the person is all the more welcome. As if to correct for the impersonality of statistics, some researchers following up on the PCPT have devised an equation that "allows an individualized assessment of prostate cancer risk and risk of high-grade disease for men who undergo a prostate biopsy." Taking into account age, race, background, PSA, and other variables, the formula, though necessarily reductive, enables a patient to calculate risk in a manner more responsive to his own history than more generic methods. This I take to be one of the more promising directions of prostate research, introducing a ray of clarity into the obscurity that engulfs prostate medicine, and implicitly recognizing the patient as a unique being, a person. As physicians screen for patients at a risk high enough to offset the risks of chemoprevention, they will probably come to use supple mathematical models.

2. Psychological Injuries

As we know, the most promising result of the PCPT—a 25 percent reduction in the incidence of prostate cancer in the finasteride group—was overshadowed by the higher incidence of aggressive malignancies in the same group. Statistical benefits made less impression than clear and present dangers. And the different rhetorical weight ascribed to these outcomes put the defenders of finasteride at a disadvantage. In a pointed reply to an article in the March 2006 issue of the *Journal of Urology* that reviews the finasteride data and counsels caution, some of the authors of the original *New England Journal of Medicine* paper voiced their dismay that too little attention was paid to the 25 percent figure and its significance. "Given that the PCPT was not designed to assess grade and that biases in ascertainment [that is, detection] most likely affected the outcome," they wrote, "we are surprised by the lack of focus on its primary end point [that is, the event being measured]—*diagnosis of prostate cancer*—with a 24.8% risk reduction for finasteride." They continue:

> Omitted from the discussion is that in [a] 1,000 subject cohort, 60 men would be spared the diagnosis of cancer. Given that more than 90% of newly diagnosed men receive definitive treatment, and suffer the stigma, anxiety and side effects associated with diagnosis and treatment, omitting this benefit from analysis ignores the real impact of a prostate cancer diagnosis in the United States today.

(The more frequently cited figure, indeed the figure originally released by the directors of the PCPT, is 15 cancers prevented per 1,000 men per seven years.) It is as if finasteride's defenders, frustrated by the other side's rhetorical advantages, sought to even the score by equipping 60 hypothetical men with a capacity for psychological pain. The 60 seem to suffer as much from the diagnosis of cancer as from cancer itself. They were indeed at risk of diagnosis.

As psychology is loosely allied in the public imagination with medicine (thus sharing in the good repute of medicine by association), so in this instance, ironically enough, the advocates of finasteride seek the assistance of psychology to strengthen their case. Like lawyers making overblown claims of mental pain and suffering, the advocates of finasteride inflate their estimate of prostate cancer's harms under the guise of assessing the real impact of the disease. Arguing that prostate cancer patients suffer the equivalent of psychological trauma, they follow after pop psychologists who console readers for their injuries and offer release from the oppression of vicious labels and the bondage of poor self-esteem. Martin Luther King, Jr., wrote the "Letter from Birmingham Jail" on behalf of a stigmatized people afflicted with anxiety—"plagued with inner fears," as he put it. To put prostate cancer patients in the same category as an oppressed people, or even a similar category, constitutes an abuse of rhetoric. While it is true that the anxiety caused by cancer screening "should not be underestimated" (in the words of another paper already cited), this patient can report from experience that prostate cancer does not stigmatize. That it is hard to come to terms with cancer, to acknowledge it, to label oneself a cancer patient, does not mean that that arch-villain, "society," affixes a stigma to you and treats you with contempt. King's letter appeared in a collection entitled *Why We Can't Wait*, and the real connection, it seems to me, between prostate cancer and civil rights is not that we patients have been dealt with unfairly but simply that the finasteride debate began before time established whether or not the finasteride side of the PCPT really had worse cancers—such was the general impatience. The self-evident point that any patient would prefer not having cancer in the first place can be made without invoking stigma.

> A final, often-forgotten aspect of chemoprevention is the effect on the individual; although most individuals will express a desire for early cancer detection and treatment, a preferred alternative is the opportunity to never have to face the diagnosis in the first place (i. e., prevention).

Fair enough.

In using the soft-science idiom of psychology, physicians seek, in effect, to win back the affections of patients who began to consult psychotherapists in large numbers after World War II, when medicine became far more powerful but far less avuncular and consoling. But why would finasteride researchers inflate the harms of a disease already harmful enough? They do so, I believe, to reclaim the language of harm from those who oppose the preventive use of finasteride on the ground that it is too dangerous. "We too respect harm," they seem to say, "but we take account of harms that you ignore." That both sides in the debate appeal to harm suggests the power still wielded by considerations of harm in medical thinking.

3. Sexual Effects

Because the risk of high-grade malignancies is a prohibitive liability, it occupies the spotlight of the finasteride literature. But it is not the drug's only liability. Another is duly noted, if passed over lightly, in the literature: adverse sexual effects, principally impotence and loss of libido. Beyond mentioning their existence, the finasteride papers pay little attention to these depressing realities, perhaps because they are things one can live with, as well as being reversible, in contrast to aggressive cancer. If the cloud of doubt hanging over finasteride were somehow cleared, doctors would probably not be dissuaded from prescribing the drug by its sexual effects, but men might not want to take it for those reasons, especially during the decades of prostate cancer's latency when the disease is ripe for prevention.

As it happens, although men flocked to the PCPT when it began recruiting subjects (an enrollment predicted to take three years took only seven months), a good many ceased taking their medication once enrolled: 14.7 percent on the finasteride side and 10.8 percent on the placebo side. My guess is that when these men, aged fifty-five and older, noted a decline in their sexual energy, they thought their pill might have something to do with it and left it in the bottle. Most of them got the pill right. The difference in the "rates of nonadherence to medication," as they are called, is significant enough that some think it helped skew the data of the PCPT against finasteride.

A director of the PCPT once remarked in an interview that the ideal candidate for finasteride would be a patient "who is concerned about the development of prostate cancer, has a higher-than-average risk of the disease, has urinary symptoms that may be relieved by finasteride, and is not sexually active." One wonders how many men who might be interested in arresting prostate cancer before it gets started, which means getting it early, fit this Platonic profile. An economic analysis of the finasteride question presumes a 50-year-old man who takes the pill for twenty years. Is he celibate? Considering that "approximately 30% of men between the ages of 30 and 50 years have histologic cancers in their prostate," some might say that prevention should begin earlier than 50. How many would risk the forfeit of their sexuality for the chance of being spared an overtreated disease some decades later? The medical literature ignores this question while the press, which might have been expected to pick up on it, largely missed it.

In a brief notice—hardly a story—on the PCPT published in July 2003, a *Time* writer reported that

> compared with a placebo, Merck's finasteride . . . appeared to reduce prostate cancer by 25%. But in the seven-year study, involving more than 9,000 men ages 55 and older, the finasteride group . . . had a slightly higher rate of aggressive, "high grade" tumors, which are harder to treat. Complicating matters, finasteride also caused loss of libido and impotence.

Figured as a percentage like the reduction rate, the increase in aggressive disease is not slight but marked. The sexual side effects, too, show up in percentages of men, not (as the word "caused" may imply) in everyone. Devoting one sentence each to medical dangers and sexual woes, as if the two somehow weighed equally, may also seem careless, but at least the writer gives the latter issue its due, unlike a medical literature concerned almost exclusively with medical consequences. For some men, the prospect of sexual side effects might rule out finasteride even if it were quite safe. For some, the sacrifice of sexuality in the interest of reducing risk would represent a good deal more than an ounce of prevention. Some, wanting things both ways, will take finasteride with one hand and Viagra with the other. Some will be content to leave the calculation of risks to actuaries. "Sufficient unto the day is the evil thereof." Some will say they are not such self-deniers, such ascetics, as to court loss of libido and impotence for the sake of some theoretical benefit. For some, writing off the body in the name of health will make no sense. Some just will not adopt a save-your-pennies approach to life.

People may or may not act in their own rational interests, but is it clear that neutralizing your own androgens is even rational? Regardless of numerical projections, men simply cannot be expected to behave uniformly, categorically, like beings who will do anything to reduce risk, all the more if the risk in question isn't that clear and compelling to begin with. The willingness of men by the millions to follow medical advice cannot be assumed. Just as an appreciable number of the men enrolled in the PCPT failed or declined to take the assigned medication, so too did a great many refuse the "offer" of a final biopsy. When the PCPT was being designed in 1992, the architects of the experiment anticipated that only 5 percent of the participants would refuse, and a subgroup of the steering committee of the PCPT made it its business "to ensure that participants remained involved in the trial and were committed to getting their end-of-study biopsy." Nevertheless, 3,927 men out of 16,295 who completed the study—close to 25 percent—refused. Anyone who has endured this procedure will understand why.

A few years ago, before the report of the PCPT, a wry spirit observed in the British medical journal *The Lancet* that finasteride "acts as an antiandrogen, and consequently can slow development of male pattern baldness," while testosterone injections, pills, and gels fight "andropause" but promote baldness.

> If present trends continue, sometime in early October, 2004 half of middle-aged male narcissists will be on finasteride and the other half on testosterone. The half on testosterone will be bald, and will be asking their doctors about finasteride. The half on finasteride will have problems with muscle strength and libido, and will be asking for testosterone. Then we will witness the great clash of molecules.

The writer's point being that you can't have everything. "Modern life is reminiscent of a class of fairy tales in which the genie grants three wishes, but the recipient always manages to mess it up by over-reaching, by wishing for too much." It bears noting, though, that the dosage of finasteride which in this little fable is enough to send men scurrying in search of testosterone is a fraction of the five milligrams that reduced prostate cancer by 25 percent in the PCPT. Researchers who project the general use of finasteride— who presume, that is, that millions of men will consent to risk its sexual effects—may be weaving a fairy tale of their own.

The narcissists in the parable, taking a 5 alpha-reductase inhibitor for the sake of their hair, might be dismayed to learn that a congenital deficiency of that enzyme is associated with pseudo-hermaphroditism. Males who inherit a 5 alpha-reductase deficiency are born with ambiguous genitalia, though they do keep their hair (as do eunuchs). Recall that they also have a rudimentary prostate; that it was the discovery of this connection in the mid-1970s that got researchers thinking about a drug to shrink the prostate by mimicking the same deficiency; and that it was the resulting drug's efficacy that got researchers thinking about its potential to prevent prostate cancer. Traces of this story, which begins with the study of males who as children resemble females, persist in the warning to pregnant women against even touching a broken finasteride pill because of a certain unspecified "potential risk to a male fetus." (There may in fact be a lot of broken finasteride pills about. Reports tell of men breaking their five milligram pills of Proscar into fifths to get around the inflated price of one milligram pills of Propecia.) Presumably for the same reason, when Proscar was originally approved by the FDA in 1992, women who were pregnant or intended to become pregnant were warned against exposure to the semen of men taking the drug, even though finasteride was present only in minute amounts. Would men who are already somewhat neurotic about their masculinity knowingly take a drug with this drug's history?

The shadow of sexual impairment and loss hangs over the medical literature of prostate cancer. One of the finasteride authors has also studied castrates in China, the Ottoman Empire, and a fanatical Russian sect. His interest is not amateur but professional. "Trials of orchiectomy"—that is, castration—"were done in the 1940s on men with metastatic prostate cancer with evidence of . . . improvement," a finding considered the point of origin of hormone therapy for prostate cancer. Nowadays testosterone is reduced chemically "to castrate levels" in men with advanced prostate cancer. Here and there the medical literature reminds us that eunuchs are free of prostate cancer, as if in this respect at least they had luck. A number of men on finasteride, and certain men undergoing treatment for prostate cancer itself, grow breasts, a condition referred to as gynecomastia. Indeed, in the PCPT finasteride increased gynecomastia by virtually the same percentage by which it reduced BPH, and it is a treatment for BPH. In this world of breasted men and men surgically or chemically mutilated, those born with ambiguous genitalia as a result of a 5 alpha-reductase deficiency no longer seem quite so out of place. And like eunuchs, these pseudo-hermaphrodites do not get prostate cancer—a reminder that freedom from risk comes at a price.

[5]

Calculation of Harm

◦O◦O◦O◦

1. JEREMY BENTHAM AND THE ART OF HEALING

If one had to identify a date when numbers made a definitive entrance into medical thinking, it might be 1825, the year that saw the publication of a study of phthisis, or tuberculosis, by Charles Louis (1787–1872). The author describes his numerical method thus:

> After having grouped my cases in respect of their outward analogies, I inquired how many times any given morbid change or symptom existed in each group. In a word, I counted the number of instances in which those symptoms of anatomical changes had occurred, in order to determine their true value; for symptoms or lesions which present themselves invariably in a given disease, are of vast importance, and become more and more insignificant in proportion as they occur less frequently.

Louis designates his method as "a science of observation and of observation purely." One of his English contemporaries, the utilitarian reformer Jeremy Bentham (1748–1832), famously dreamed of employing pure observation in a quite different sense—as a control mechanism in a penitentiary. He too, however, appealed to numerical standards, most notably via the principle of "the greatest happiness of the greatest number," which he popularized.

Although himself interested in medicine, Bentham sought to practice his corrective art not on this person or that but on a disordered world. As he once wrote,

> The art of legislation is but the art of healing practiced upon a large scale. It is the common endeavour of both to relieve men from the miseries of life. But the physician relieves them one by one: the legislator by millions at a time.

In germ here is the distinction between effects on persons and effects on categories at issue in the finasteride debate. Such was the scope of Bentham's ambitions that he dreamed of designing constitutions and legal codes, creating a Ministry of Health, and reforming the criminal and the poor. Significantly, the prison he sought to build (where inmates would be kept in check by continual surveillance) was itself intended as a wholesome, sanitary institution, a device to "'cure' the defective mechanism of the human frame and the human mind," in effect to do medicine on a large scale. While his more visionary ambitions did not come to much, followers of Bentham did apply themselves to the improvement of public health, and his bold critique of existing institutions in the interest of the greatest good of the greatest number gave rise to an influential movement, just as his way of analyzing, reducing, and calculating in the name of enlightened rationality prefigures the modern social sciences. Bentham computed the rightness of a course of action arithmetically, by weighing its projected pleasures against its pains, both broken down by intensity, duration, certainty, and so on. "The heavier side of the scale determined the action." Descendants of Bentham's scale are everywhere in the finasteride papers.

Those today who consider the doctor-patient relationship a "remnant of the old nobility of medicine" unwittingly take after this man who mocked the archaism of British institutions and hoped to see the aristocracy's monopoly on power broken, even as he sought to practice the art of healing on an unprecedented scale. Echoes of Bentham are also heard in the medical debate over the Hippocratic rule, with critics deriding it as a bit of sonorous nonsense much as Bentham derided the shibboleths employed in his time to obstruct progress. One such critic sketches a scenario in which medical students and a senior doctor stand

> at the foot of a patient's bed discussing possible courses of action for the patient. Tentatively, one of the students suggests a therapy, at which point the house officer raises his forefinger upward and in a profound voice says, "Ah, but *Primum non nocere*: First, do no harm." A calm, wise, satisfied smile appears on his face as all the students reflect on the weighty import of his words. As the students then walk out on the way to class, each is inspired and awed by the wisdom and nobleness of his or her chosen profession.

Bentham regarded British institutions themselves as a kind of invalid whose treatment was being thwarted by preposterous arguments that nevertheless sounded weighty and wise. Foremost among the fallacies employed to block reform was the appeal to the wisdom and nobility of ancestors, or what Bentham labeled the Chinese Argument. "Our wise ancestors—the wisdom of our ancestors—the wisdom of ages—venerable antiquity—wisdom of old times—such are the leading terms and phrases" of arguments that he thought served only to keep an ailing body politic ailing. Both Bentham and the critic of the Hippocratic maxim wield mockery like a scalpel, and both debunk the watchwords and idols of the past. That Bentham donated his body for dissection reveals both his disdain for the age-old taboo on that practice and his esteem for medical science grounded in anatomy. (Charles Louis collected much of his data from autopsies.) By a poetic coincidence, the Anatomy Act legalizing dissection was passed in 1832, the year of Bentham's death.

2. THOU SHALT NOT

Critics of Bentham in his own age noted his reduction of judgment to a species of computation. His illustrious disciple and godson John Stuart Mill, who eventually departed from the master's teachings, described Bentham's pursuit of exactness thus:

> That murder, incendiarism, robbery, are mischievous actions, he will not take for granted, without proof. Let the thing appear ever so self-evident, he will know the why and the how of it with the last degree of precision; he will distinguish all the different mischiefs of a crime, whether of the *first*, the *second*, or the *third* order; namely, 1. The evils to the sufferer, and to his personal connections; 2. The *danger* from example, and the *alarm* or painful feeling of insecurity; and, 3. The discouragement to industry and useful pursuits arising from the *alarm*, and the trouble and resources which must be expended in warding off the *danger*. After this enumeration, he will prove, from the laws of human feeling, that even the first of these evils, the sufferings of the immediate victim, will, on the average, greatly outweigh the pleasure reaped by the offender; much more when all the other evils are taken into account. Unless this could be proved, he would account the infliction of punishment unwarrantable.

To that length Bentham will go in order not to say, "Thou shalt not kill," "Thou shalt not steal." While his way of dissecting questions has little to do with medicine (whatever his fantasy of ministering therapeutically to the body politic), medical papers still tabulate and enumerate, still weigh sets of numbers against one another. Where Bentham does not say, "Thou shalt not kill," the author of a paper on the finasteride question does not actually invoke the first principle of his profession, Do No Harm; but perhaps because medicine, being a tradition, is more deeply attached to the past than Bentham was (for Bentham "begins all his inquiries by supposing nothing to be known on the subject"), medical thinking on the finasteride question continues to be informed and constrained by the principle of not doing harm, even when critics deride it as so much antiquarianism, and even when the computation of benefits might justify harm.

That benefits properly computed might well overbalance the assumed risks of finasteride appears to be the message that some of the authors of the original finasteride study in the *New England Journal of Medicine* wish to send their critics. Consider again their "editorial comment" on a cautious analysis of finasteride and the prevention of prostate cancer. While acknowledging the "mixed outcomes of the PCPT," they are disappointed that its finding of a 24.8 percent risk reduction for the finasteride group—the number the study was designed to ascertain—has slipped from the center of the discussion. They point out that even their critics have found that for a group of 1,000 men, "a net increase of 140 person-years accrues from finasteride," and they remind the critics that of these 1,000 men, 60 "would have been spared the diagnosis of cancer" and its accompanying miseries. (Again, the official NIH figure is not 60 but 15.) In effect the drug's defenders argue that regardless of the question of high-grade tumors, the numbers in favor of finasteride are compelling enough to warrant consideration of its use for purposes of prevention. The prohibition against doing harm is finessed by a show of numbers (and, as I suggest, an inflated estimate of psychological injury).

If indeed a case can be made that finasteride would benefit society in the form of cancers prevented or person-years saved, more than it would cost in the form of cancers aggravated—as some argue even now—only an aversion to causing harm stands in the way of the medical profession's embrace of the general, population-wide use of finasteride as a chemopreventive. In the spirit of Bentham's disbelief in the idols of tradition, many consider that aversion a hangover from the past. Recall that according to a critic of Hippocratic medicine cited earlier, the pursuit of "the maximum benefit for the greatest number" reduces the principle of avoiding harm to an archaism. But if the crux of medicine is the consultation between patient and doctor, exactly how is the doctor to inform the patient that he considers fidelity to the precept Do No Harm a piece of "pointless nostalgia" for an era when doctors could afford that ethical nicety? How is he to convey that his foremost concern is not the welfare of the patient who stands before him at all, but that of an anonymous entity called society? Patients do not regard themselves as stand-ins for some larger entity. It is reported that when physicians carry out controlled clinical trials, the study subjects sometimes find it hard to grasp that the research is being conducted "for the good of society, and not necessarily for their personal benefit." A naive error but a very understandable one, both because the study goes through the motions of caring for the subject and because no one regards his or her own body as a surrogate of "society."

Even assuming a doctor professed himself to be a utilitarian and made it known to the patient that he placed the general good above that of the patient, communication between the two would arguably break down from that point. As the patient I

> could not get him to promise, in the style of the Hippocratic Oath, always and only to deploy his skills to my advantage; nor could I usefully ask him to disclose his intentions. . . . If I ask him what his intentions are, he will answer truthfully only if he judges it best on the whole to do so; knowing that, I will not unqualifiedly believe him; and knowing *that*, he will realize that, since I will not do so, it will matter that much less if he professes intentions that he does not actually have. And so on, until my asking and his answering become a pure waste of breath.

Perhaps communication would be less of a mockery if the doctor respected the Hippocratic Oath.

This absurdist parable dramatizes how poorly "simple Utilitarianism" comports with the practice of medicine. If only because medicine came into being long before utilitarianism did, and in its original forms had little or nothing to do with promoting the happiness of society, the marriage of medicine and utilitarian thinking is bound to be strained. A disciple of Bentham in our own time has argued that

> When the death of a defective infant will lead to the birth of another infant with better prospects for a happy life, the total amount of happiness will be greater if the defective infant is killed. The loss of a happy life for the first infant is outweighed by the gain of a happier life for the second.

If Bentham himself would never have entertained this grotesque conclusion worthy of Swift's "Modest Proposal," it is because he was deterred by some lingering regard for tradition after all—specifically, the traditional interdiction of killing. Perhaps, however, Bentham or his followers would have gone along with the physician who, in the 1950s, defended the existing practice of hospital medicine despite the occurrence of life-threatening or fatal complications in 5 percent of all hospitalized patients. Regardless of the interdiction on taking life, the physician in question regarded such casualties as "the price we must pay" for medical progress—in effect, as so many sacrifices to the greatest good for the greatest number.

In spite of the universality of risk/benefit language in medicine, the model of risks and benefits on opposite pans of a balance scale does not necessarily suit medicine—especially preventive medicine—because of the special character of medical risks and the higher standards that bind medical practice. A nineteen-fold return makes for a very handsome investment (and the PCPT itself, which cost some $73 million, has been portrayed as an investment), but five deaths for every ninety-five surviving patients do not make for good medicine. While as a matter of utilitarian policy it may seem to make sense, the policy of sacrificing the welfare of a few to the welfare of many is potentially in contradiction with the ends of medicine. Finasteride has not been blamed for any deaths (the deaths due to prostate cancer being very few on each side of the PCPT), but if, in the present state of knowledge, it were administered generally, not only would the welfare of those visited with more aggressive cancer be sacrificed to a larger group—in this case a few times larger—but the benefits enjoyed by that larger group would be nebulous, owing to the inscrutable nature of prostate cancer itself, a disease known to be overtreated.

The principal point of conflict between utilitarianism and the traditional practice of medicine is surely that the former, but not the latter, accounts the welfare of the patient, or any single person, secondary to a larger consideration. If my doctor is a recognized utilitarian, "I know, of course, that his intentions are generally beneficent, but equally that they are not *uniquely* beneficent toward me." In the spirit of such enlarged thinking, some seek to broaden the scale of medical concern to the public or population level. Bentham himself in effect broadened the scale of Lockean thinking, applying some of the educational principles of Locke—a physician—to entire categories of people. As laid out in his *Thoughts Concerning Education*, Locke's pedagogy calls for gentle methods, strict oversight, and careful attention to the formation of habits, the object of all this solicitude being a single child. Bentham, like other paternalistically minded reformers, adapted these guiding principles to a field of application very far from the original context of a gentleman's household. It was Bentham's ambition to reform the habits and character of any number (and these of the criminal and the poor, not gentlemen) by means of the surveillance mechanisms and wholesome regimen of his ideal prison, the Panopticon. More of the Panopticon shortly; for now it is enough to say that this, Bentham's most renowned or notorious project, was intended to subject thousands to something like Locke's methodical, enlightened, medically informed method of education, originally designed for a single person. If criminals were, as Bentham believed, like children, why couldn't educational methods conceived for a child be adapted to this group? While the modest scope of Locke's design was lost in this process of scaling up, Bentham's ambition of extending the benefits of character formation to entire groups was consistent with his emphasis on "the greatest number." A founder of the Statistics Society, he might also be considered an honorary forerunner of the sort of statistical tabulation that is the very currency of debate in medical literature, including the finasteride papers.

So committed was Bentham to utilitarian computation that he thought it could solve even difficult moral problems. If the greatest happiness of the greatest number is paramount, then couldn't a majority oppress a small minority, reasoning that the keen pleasure it derives from persecuting its neighbors increases the general stock of happiness, even after an adjustment is made for the minority's pain? (In and of itself, John Stuart Mill's eloquent critique of the tyranny of the majority in his essay *On Liberty* points to his break with Benthamism.) Bentham, however, thought that computation would prove the majority wrong—that numbers in this case represent not the problem but the answer. "He believed that it would be possible for a minority to be oppressed by a majority in a way which caused more unhappiness to the former than it brought happiness to the latter, and which therefore reduced the overall happiness of the community." Evidently much depends on the numerical weight assigned to this or that pleasure or pain. A medical disciple of Bentham might argue that even if finasteride prevented more cancers than it aggravates, the smaller number of high-grade cancers nevertheless cause so much distress as to overbalance the gains of prevention (by analogy with the few whose happiness is unjustly sacrificed to the many). Perhaps some of finasteride's opponents do reason in this way. But the numbers game is a treacherous one, and if the finasteride issue were to be decided today by computation, it would probably be decided in favor of the general use of the drug, not against it.

3. The Weight of Pain

The idea of weighing harms is not so unknown to humankind that we have to confront the issue as if we were the first on earth to do so. The law does it. Three years before Bentham's birth, his countryman Samuel Johnson argued powerfully that it makes no sense to punish both robbery and murder—crimes "very different in their degrees of enormity"—with death. To equate these crimes is "to confound in common minds the gradations of iniquity, and incite the commission of a greater crime to prevent the detection of a less." What to Johnson seemed an irrationality in the law would have seemed to Bentham one more proof of a thoroughly gothic institution. In the face of such medievalism, Bentham set out to reconstruct not only law codes but houses and practices of punishment, almost as if he really were the first on earth to lend his thought systematically to these things. Today some physicians grapple with the problem of weighing medical outcomes as if they too could receive no guidance from the past. But setting aggressive cancers categorically apart as grievous harms should not be so much more difficult than distinguishing "degrees of enormity" or "gradations of iniquity" and placing murder in a separate category from lesser offenses, as in fact we now do.

The question of how to weigh pain, or harm, vexes physicians and researchers concerned to measure the value of finasteride (but also its sister drug, tamoxifen) as a chemopreventive. Confronted by the mixed results of the PCPT, a research team sought

> some means by which to compare the adverse impact of 48 low/intermediate grade prostate cancers plus 12 high-grade cancers versus 27 low/intermediate grade prostate cancers plus 18 high-grade prostate cancers, and 15 cases prevented [in a group of 1,000 men].

At first they assigned all prevented cases of cancer a value of zero, and all cases of cancer one, which yielded the conclusion "that the prevention is beneficial." This minimalist model does not reflect the difference between high- and low-grade disease, however, and so the researchers tried an alternative in which high-grade cases were given a weight of two. This too yielded the finding that prevention is beneficial, if less so. The absurdity of both formulas being painfully obvious, the researchers gave up assigning numerical weights to medical outcomes and instead compared the ten-year survival rates of fifty-five-year-old men with low-grade prostate cancer against those with high-grade disease, and both of these groups against the ten-year survival data of men in the general population. A model in which everyone who survives receives a value of one (a sort of medical one man, one vote) does away with the problem of weighting different degrees of disease. By this means the researchers were able to compute the person-years that would be saved as a result of the preventive use of finasteride—a total which varies depending on the assumptions about finasteride built into the model, but which definitely vindicates the use of the drug by millions of men.

"The person-years model shows that more than 300,000 person years would be saved during a period of 10 years with the widespread use of finasteride, assuming no change in the rate of high-grade prostate cancers." Because the existing finasteride data do not allow that bold assumption, the authors factor in elevated rates of high-grade disease, concluding that the rate would have to triple in order to cancel out the survival benefits of finasteride. "The potential detrimental effects of an increased rate of patients who have prostate cancer with high-grade Gleason scores would be outweighed by a reduction in incidence." Here, then, is a medical analogue of the case in which the total happiness in a community is increased at the expense of some minority of its members. In the mid-twentieth century many in American medicine were so confident of the mission and power of medicine that they were willing to write off incidental harms to persons as so many side effects of progress. Utilitarians in effect if not in name, they believed that medicine's contribution "to the health of the United States citizenry and, indeed, of humanity supplied the implicit justification for many medical harms to individuals." So in this case. The reduction of medical judgment to a numbers game, and the conception of medicine itself as a means to advance the greatest happiness of the greatest number (both, it seems to me, in the spirit of utilitarianism), dissipate the cloud of doubt surrounding finasteride and make the virtues of its general use seem not only obvious but indisputable. The numbers may be indisputable, but they are also disembodied and remote from reality. When the authors argue that

> even if we hypothesize more than a doubling of the proportion of high-grade tumors in the cancer population (from 19.2% to 39.2%), 159,680 person-years still would have been saved, representing a positive benefit to society,

their analysis reads unintentionally like a parody of reason. Seeking the last degree of precision, the authors achieve only surrealism.

Among urologists practicing medicine at ground level, there must be very few who would decide to use finasteride even if its risks were worse than the PCPT numbers indicate. Precisely because to the medical profession the risks of the drug are not a matter of indifference, most of the medical debate concerns whether the risks are real or only apparent, not what benefits the drug might bestow on some sector of the male population of the United States even if it were twice as dangerous as the PCPT suggests. For most practitioners of medicine, the risks of finasteride revealed in the PCPT are quite troubling enough, without being doubled. The ease with which the just-quoted authors make unthinkable assumptions and defy what I have called the magnification effect, and the confidence with which they propose a course that few conscientious practitioners would recommend to a patient, point to the sort of medicine that might be practiced if physicians were competent and careful but out of touch with the prohibition of harm. Only physicians so devoted to the principle of the greatest good for the greatest number that they could write off untold suffering caused in its name—only these could contemplate without flinching a doubling of those malignancies that a cancer patient himself would most dread.

4. THE IDEAL OF PREVENTION

Once the utilitarian standard of the greatest good for the greatest number is in use, the question arises, What about the lesser number? We are not the first generation to wonder whether harm to some can be offset by benefits to others. The issue was posed in the medical literature forty years ago by authors who questioned "whether the good derived by humankind from advances in medical science justified harm to particular patients—that is, whether the utilitarian rationale for medical harm was still valid." At the time, however, the discussion of medically induced harms focused on the risks of treatment, such as the dangers of anesthesia or the careless use of antibiotics. With finasteride matters have taken a postmodern turn, for now the issue is not whether the benefits of medical *treatment* outweigh the harms, but whether the more speculative benefit of *preventing* (that is, reducing the incidence of) an overtreated disease justifies making the disease itself worse in a minority of cases.

Even in the instance of vaccination, which represents the ideal of prevention, weighing harms and benefits is not as simple a matter as the reassuringly familiar image of the balance scale implies. Regarding polio vaccination in India, it is difficult to arrive at good estimates of both cases prevented and cases caused, but that there *are* cases of "vaccine-associated paralytic poliomyelitis" (VAPP) is troubling in itself. A thoughtful analysis of "Some Ethical Issues Arising from Polio Eradication Programmes in India" brings out the difficulty of reconciling the moral books in any such large-scale program:

> All vaccinations carry some risk of harmful side-effects. While such risks are generally very low, even very low risks are important to take into account in any assessment, for two reasons. First, affected individuals will be indifferent to the statistical size of the risk because the event occurred, and occurred to *them*, and second, given the millions of individuals vaccinated, the actual numbers affected by even a low probability side-effect may be quite large. This is certainly the case when the target is to eradicate a disease requiring the vaccination of millions of people.

Similar considerations would come into play if finasteride were used preventively by millions of men. Assuming as we must that the drug's risks are real, more aggressive cancers would flow from its use at a certain rate (a much higher rate than that of vaccine-associated polio), and those on whom these consequences fell, though they could not trace them to the drug with any certainty, would certainly not write them off cheerfully as side effects of progress. The more finasteride takers, the more casualties there would be. If the ready-made model of the balance scale seems unequal to the issues of polio vaccination (and polio is both infectious and eradicable, unlike prostate cancer), neither will it really adjudicate the finasteride question.

The comparison of risks and benefits evolved from the utilitarian balance scale for weighing pleasures and pains employed by Bentham, among others. (Carlyle, railing against Benthamism, once wrote, "Man . . . has fancied himself to be most things, down even to an animated heap of Glass; but to fancy himself a dead Iron-Balance for weighing Pains and Pleasures on was reserved for this his latter era.") And like medical researchers today, Bentham too was concerned with prevention—the prevention not only of disease (through vaccination) but fires, floods, famines. Indeed, he has been cited as "the direct begetter . . . of the science of preventive medicine," in that his ideas inspired the Public Health Act of 1848. A well-designed government, Bentham thought, should include a bureau of prevention. Intended as it was to exclude filth—understood at the time as the cause of disease—and preclude exposure to the sources of moral infection, the visionary prison for which Bentham is best remembered today might be interpreted as a sort of elaborate preventive device. Its distinguishing features—the confinement of prisoners in cells under the watchful eye of a warden; rigid, virtually medical discipline; the exclusion of brutality—all attest a concern to keep out the disorder and corruption of the traditional jail. (In the 1870s British hospitals began to be cleaned up in the way Bentham would have liked to clean up jails.) In the spirit of prevention, both tobacco and alcohol would be prohibited in Bentham's prison, and inmates would be kept "very safe and very quiet" in their well-lit cells, protected from cold, damp, and other dangers to health. As paradoxical as it may seem, such a house of correction thus bears a certain resemblance to utopias sequestered from the world, removed from the sources of contagion. The best-informed historian on the subject in fact places Bentham's design for the Panopticon in the utopian tradition. Some of the fantastic commonwealths we call utopias are as inimical to human liberty in their own way as Bentham's Platonic prison.

Bentham liked the word "liberty" less than "security," and many of those who dream of a comprehensive regime of prevention seem to lean the same way. After all, in order for such a regime to work it would have to be backed up by a great deal of manipulation, conditioning, and indoctrination, all of which argue a poor opinion of the other person and of liberty itself. If the health of the people is the supreme law ("salus populi suprema lex"), as Bentham among many others seems to have believed, the supreme law overrides all other considerations and justifies doing things to people for their own supposed good. Some who hold that issues of public health "dominate the world," and that it is the "mission" of medicine to regulate human beings in their own interest, think so little of liberty that it scarcely enters into their discussions except as a kind of nuisance factor. (One such advocate of public health cites incidentally the control of typhus by means of "proper isolation of the sick" without actually mentioning the medical reign of terror that accompanied such measures.) Distressed at the volume of manipulative ads encouraging people to drink more and drive faster cars, a noted epidemiologist reluctantly concludes that physicians ought to counteract these messages with manipulations of their own. "Maybe freedom suffers less if it is attacked from both sides." These words are ominous.

The shade of Bentham flickers through the language of the finasteride papers whenever they seem to look down on the human scene from above. The novelty, and most notorious feature, of Bentham's Panopticon is the surveillance of the inmates by a warden centrally placed to keep watch on them all. Ironically, the federal registry tracking the incidence and outcome of cancer in some 10 percent of the American population—the body of information that allows the computation of person-years lost to cancer or gained by prevention—is known by the acronym SEER: Surveillance, Epidemiology and End Results. Drawing upon this registry, the researchers who maintain that the benefits of finasteride statistically dwarf its risks build varying assumptions into their calculations: that the risk of high-grade cancer increases 6.9 percent or 8.9 percent with finasteride, that it doubles, that it triples. But they also build into their model a far larger assumption, one they don't mention and may not even have considered: that multitudes of men can be coaxed into taking finasteride indefinitely, despite its risks, costs, and sexual side effects. Only if multitudes first agreed to take the drug would "300,000 person-years . . . be saved during a period of 10 years . . . assuming no change in the rate of high-grade prostate cancers," and a still-significant number of person-years even if the rate doubled. That the researchers simply take this critical assumption for granted bespeaks a certain disdain for human liberty. How indeed could chemoprevention go into effect "at the population level"—that is, how could millions of men arrive at the same decision on a profoundly ambiguous and vexing question—unless they were steered into it?

Now and then in the finasteride papers a note of heavy-handedness creeps in. Sometimes the researchers make a kind of unconscious assumption that the doctor takes the decision for the patient, as when we read that in some cases the numbers "may reasonably translate into a decision by the physician to give the drug for cancer prevention and then to watch the patient carefully." Sometimes uncompliant research subjects are simply edited out. "After 7 years of [the PCPT], all subjects with a history of normal screening examinations were given a biopsy of the prostate," writes one author, ignoring all who refused. While two researchers are quite right to conclude that "placing all black men on 5ARIs [that is, 5 alpha-reductase inhibitors] may be premature given the current level of evidence," reality argues that doctors do not have it in their power to put all black men on anything, even if a doctor's "prescription" still carries a lingering resonance of the word's original authoritative meaning. Given the legacy of distrust left by the Tuskegee Syphilis Experiment, any attempt to "place all black men on 5ARIs" would be interpreted as wholesale chemical castration. Patients have a mind of their own. Tamoxifen is not in general use because it is subject to the double constraint that only so many physicians will recommend it, and only so many patients will accept it.

The doctor who introduced the obstetric use of chloroform, James Young Simpson, taunted his opponents by saying that no matter how strenuously they opposed anesthesia, "our patients themselves will force the use of it upon the profession. The whole question is, even now, one merely of time." Women giving birth could not be expected to accede to "the conservatism of the doctrine of the desirability of pain." (Bentham rejected such a doctrine axiomatically and would have had no truck with biblical justifications of pain in childbirth.) But even the most committed defenders of finasteride cannot say that their reluctant colleagues will end up writing prescriptions for the drug willy-nilly because patients will demand it.

As Isaiah Berlin, the great historian of ideas, has written, the early utilitarians, including Bentham, believed in molding and manipulating others for their own benevolent ends (the Panopticon itself being intended to do exactly this). But "to use [others] as means for my, not their own, independently conceived ends, even if it is for their own benefit," is to degrade these others, "to behave as if their ends are less ultimate and sacred than my own." These are words researchers ought to engrave on their hearts as they ponder schemes to induce millions to do things for their own supposed benefit.

5. Models and Scale

"The art of legislation is but the art of healing practiced upon a large scale," claimed Bentham in a leap of metaphor, anticipating the trend toward large-scale medicine. But there are diseconomies as well as economies of scale. A quite different and more traditional ideal of healing envisions the care of many without loss of the identity of each. I am thinking of the pastoral model, in which a man of God, a curate, cares for a flock—the words "cure" and "care" being etymologically conjoined.

The Parson in Chaucer's *Canterbury Tales* is portrayed as an ideal but by no means bland figure, one who, unlike many of his fellow pilgrims, actually lives up to his duties, an example to his parish, virtuous but not haughty, neither lording it over the poor nor bowing down to the rich, a shepherd devoted to his flock.

Lest we think his dedication a mere abstraction and his flock a generality, however, we learn he will visit "the farthest in his parish" on foot, no matter the hardship—in effect making house calls—a detail distinctly suggesting that his care extends to his parishioners singly and not just collectively. If he is, as we are told, "a shepherd," a good shepherd keeps track of each and every one of his sheep (a point at the heart of a remarkable late-medieval mystery play). Some of the Parson's flock are described as poor, some as well-to-do, some as sinful, some obstinate. Here then is a model of care in which persons remain distinct beings even as they figure as a group. And some lingering echo of this model continues in medical care, much as the evocative word "care" still carries traces of its own religious history (a history also suggested in the name of the first promising synthetic drug, Salvarsan, deriving as it does from the Latin "salvare," to save, and recalling "salvation").

So too does the story of the good Samaritan continue to echo in medicine, and if the Canterbury Parson is depicted as a shepherd, his original is the teller of that story, the Good Shepherd, Christ. It bears remembering that Christ, like Socrates and for that matter Confucius, left no writings, as if he were reluctant to address something as abstract as the world. He addressed himself in the first instance to the persons he spoke to—his practice in this respect warning, perhaps, against the loss of reality that so often sets in when persons give way to abstractions, grouped by the million.

6. QUANTIFICATION AND THE AVOIDANCE OF HARM

When did American medicine begin to frame medical questions statistically? It may have been during the nineteenth-century anesthesia debate. Anesthesia lent itself particularly well to this sort of enlightened analysis, being very much in the spirit of the Enlightenment's program of reducing suffering as well as consonant with the humanitarian imperatives of the anti-slavery movement. In the thinking of Bentham and others of like mind, pain itself, after all, was evil.

While ether as well as nitrous oxide ("laughing gas") had been toyed with as a public amusement for some time, its value as an anesthetic was first demonstrated by a Boston dentist in 1846, and within months the method was in use in the foremost hospitals of New York, London, and Paris. Two years later a committee of the fledgling American Medical Association pondered the question: "Do the risks and evils attendant upon the use of [anesthetic] agents in surgery counterbalance the advantages afforded by exemption from pain, and to what extent and under what circumstances is it proper to use them?" This issue, it seemed, was best settled by calculation.

Now the *numerical method* of statistics had entered medicine, providing a means of calculating the comparative results of treatment and nontreatment. "The prudent and judicious physician . . . calculates probabilities as accurately as he can at every step, and endeavors to make every measure tell upon the great result, avoiding, as far as possible, those which will not, and especially those which will hinder or defeat it," advised Dr. Worthington Hooker. In the years following the introduction of anesthesia, many studies attempted the quantification of results that would allow physicians to do just that.

The reduction of medical judgment to a form of calculation, the weighing of risks and benefits—this has become the very language of medical deliberation. All parties to the finasteride debate speak it. Note, though, that the anesthesia debate was framed by the principle of avoiding harm, just as Hooker underscores the physician's duty to use great care, whatever his calculations. In the original, Hooker's statement is embedded in all kinds of warnings against doing harm: warnings against causing injury to patients, against curing the disease while destroying the patient, against pursuing a ruinous course of treatment, against firing shots at disease at random. Hooker's *Physician and Patient* offers a kind of epic catalogue of dangerous abuses and antics posturing as medicine. If medicine seeks to promote human well-being, the physician must religiously avoid every practice that endangers that good. Medicine is in this sense a kind of stepchild of religion itself, its rhetoric that of the highest duty.

Much as nineteenth-century medicine opened a middle way between the "heroic" course of attacking the patient for his own good and, on the other hand, the course of leaving Nature to itself, either of which could be injurious, so did physicians like Hooker and Oliver Wendell Holmes look for a middle path between the indiscriminate use of anesthesia and the refusal to use it at all. In defending the limited use of anesthesia, these exponents of what was called "rational" medicine departed from the tradition that refused to countenance risks to the patient in the interest of relieving suffering, which is not to say that they simply cast off the influence of Hippocrates. "These practitioners were willing to incur danger in order to prevent suffering, but only up to a moderate limit." In Britain only two or three deaths in a year as a result of chloroform "would have made a strong impression," while an American manual at the time cited the view that "one death in ten thousand cases is sufficient to condemn chloroform on moral grounds." Evidently the sentiment on both sides of the Atlantic was that the sacrifice of life for the sake of "merely" reducing suffering was morally indefensible. As it happens, the figure of one in ten thousand was not really an exaggeration. "That enough was known about chloroform anaesthesia to make it safe from the beginning is proved by the fact that between 1847–60, Syme, Snow and Simpson [prominent British physicians, the last of whom introduced chloroform] anaesthetized approximately 15,000 patients between them, and did not have a single fatality." Clearly, these men recognized the duty not to harm and lived up to it. Their example reflects on those who did not.

[6]

Medical Knowledge and
Medical Ignorance

"No body can have a pretence to doubt the Advice
of one, who has spent some time in the Study of
Physick, when he counsels you, not to be too forward
in making use of Physick and Physicians."—John Locke,
Some Thoughts Concerning Education

1. NATURE'S VEIL

The limited use of tamoxifen and finasteride might be interpreted as a middle way between not using them at all and using them quite generally. Perhaps if the cellular effects of these drugs were better understood, such temporizing would cease and they would be ruled in or out. As it is, practice runs ahead of knowledge, albeit it only so far. Even if the days of roving empirics calling themselves doctors are well behind us, the fact is that for all the astonishing sophistication of modern medicine, the use of both tamoxifen and finasteride remains, for now, empirical. There is much about these drugs, as well as the diseases they may "prevent" (even though they were not designed to do so), that is poorly understood, as all parties freely admit. William Osler's massive *Principles and Practice of Medicine* bears an epigraph from Hippocrates: "Experience is fallacious and judgment difficult." All contributors to the finasteride debate, whatever their leanings, recognize this principle. And being cognizant of the limits on their knowledge, they stand in the main line of the Western intellectual tradition, whatever they may think of tradition.

Consider the finasteride papers. "The high rate and *unpredictable biology* of prostate cancer make prevention of the disease an appealing strategy." "Of the cancers diagnosed in the PCPT 48% were found on end of study biopsies. Given the protracted natural history of prostate cancer and the high prevalence of latent prostate cancer in older men, *the significance of these cancers is uncertain.*" "*Theoretically*, suppression of prostatic DHT *may* also significantly inhibit the development and progression of prostate cancer." "At present, the relative importance of type 1 and type 2 5-alpha-reductase in the natural history of prostate cancer *remains unknown.* However, as dutasteride inhibits both isoforms of 5-alpha-reductase and suppresses serum DHT by more than 93%, compared with approximately 70% for finasteride, *it is possible* that greater suppression of DHT could lead to better efficacy in preventing prostate cancer." "Notwithstanding the argument that some high-grade tumors detected while the patient received finasteride may be artifactual, *there are some plausible biologic hypotheses that suggest the effect could be real.* It is likely that until this issue is settled, routine use of finasteride to prevent cancer will not be embraced widely. . . . *We are uncertain whether this question will ever be answered definitively.*" "*We do not know* the degree to which increased risk of detection of cancers [in the PCPT] resulted in underestimating the risk reduction for finasteride." "*It is uncertain whether it can ever be determined*" if the increase in high-grade tumors in the PCPT has an innocent explanation. "Some agents can be preventive and carcinogenic in the same organ. . . . [Such] problems highlight *the profound complexity of carcinogenesis.*"

"Steroid hormones, particularly androgens, are suspected to play a major role in human prostate carcinogenesis, but the precise mechanisms by which androgens affect this process and the possible involvement of estrogenic hormones are not clear." Because the PCPT evolved from the use of finasteride to treat BPH, and "understanding of the pathophysiology of BPH remains incomplete," a shortage of knowledge is virtually built into the finasteride question. It is recognized, too, that the long-term effects of a finasteride regimen are unknown—necessarily so, inasmuch as the drug itself has not been in existence very long, and went onto the market only a few years after it was first used in human subjects. All in all, the medical literature on finasteride is written in the grammar of uncertainty.

But if uncertainty is a grammar, grammars have a history. The motto of the pioneer of surgery Ambroise Paré (1510–1590), "I treat, God heals," acknowledges that for all he does and knows, there is much he cannot do and does not know. The same aphorism was quoted by Lister—author of what some consider the most momentous innovation in the history of medical practice, antisepsis—in his Inaugural Lecture at the Glasgow Infirmary in 1859. The self-consciousness of the finasteride papers, their scaled-down claims of knowledge and frank admissions of ignorance, their fine sense of ambiguity, their habitual use of qualified phrasing—all are rich deposits of the Western tradition of the search for knowledge, a search shadowed by the consciousness of ignorance. Such a cultivated sense of uncertainty as we find in the finasteride papers (an uncertainty not oppressive but enabling, not shameful but honorable) is not something that springs up spontaneously in the human soul, like a stirring of nature or an unschooled response to the world. For reasons that are surely historical themselves, many today have come to believe that everything of value in human life really does spring up spontaneously, and anything achieved is achieved in opposition to tradition.

Over the past century or so, remarks Edward Shils, there has risen in our culture a

> metaphysical dread of being encumbered by something alien to oneself. There is a belief, corresponding to a feeling, that within each human being there is an individuality, lying in potentiality, which seeks an occasion for realization but is held in the toils of the rules, beliefs, and roles which society imposes. . . . To be "true to oneself" [in this sense] means . . . discovering what is contained in the uncontaminated self, the self which has been freed from the encumbrance of accumulated knowledge, norms, and ideals handed down by previous generations.

The pop psychology movement, so influential just now, markets a hyperbolic version of this already painfully simplistic story.

Some of the finasteride researchers may express in writing more scrupulous uncertainty than they feel in their heart of hearts—not, however, because they sacrifice their authenticity to an oppressive and alien regime in conformity with the story line of pop psychology, but because they know the advance of knowledge requires the recognition of ignorance. Nor did they achieve their highly developed sense of uncertainty by casting off rules and norms, but by working within a tradition concerned to derive knowledge from the unknown. Over the same time that a strong animus against the legacy of previous generations grew up in our culture (an animus specially, almost personally, focused on the Victorians), medicine built on a foundation handed down by earlier generations. As the decisive transformation of folk medicine into modern medical science took place in the nineteenth century, physicians recognized that such country methods as bleeding and purging, and such remedies as cure-alls, accomplished nothing because those who used them didn't know what they were doing. The more modern-minded physician knew that his knowledge and power went only so far, and that it therefore behooved him to step carefully and avoid causing injury. "Even if modern doctors could not cure their patients at least an understanding of disease mechanisms and drug action *kept them from doing harm*." Avoidance of harm goes hand in hand with recognition of the limits of knowledge, an intellectual modesty nourished by a philosophical tradition going back to the age of Hippocrates.

From ancient times, the pursuit of knowledge has been bound up with the admission of ignorance. The lesson of Socrates—Hippocrates' contemporary—is

> that ignorance is a lot more prevalent than we suppose—that we don't really know as much as we think we do. Thus Socrates advises caution and the suspension of belief; we shouldn't let our strong desire to know fool us into misconstruing erroneous belief for real knowledge. . . . Socrates counseled epistemological modesty.

It is impossible to read the finasteride literature without being strongly impressed with its own epistemological modesty. The writers have taken the Socratic lesson fully to heart. *"We are uncertain whether this question will ever be answered definitively."* Perhaps if historical conditions had not favored the radicalization of doubt, Socratic uncertainty would not have entered so deeply and generally into the making of modern knowledge. In any case, the discovery of new lands, the collapse of the medieval consensus, and the outbreak of wars of religion in the early modern era had the effect, in some quarters, of throwing the very grounds of knowledge into doubt. Confronted with profound and intractable differences of opinion, some were forced "to reflect not merely about the things of the world, but about thinking itself and even here not so much about truth in itself, as about the alarming fact that the same world can appear differently to different observers"—much as writers on the finasteride question are acutely aware that the meaning of their data does not speak for itself, that others read it differently, and that the ground under their feet is contested territory. Michel de Montaigne (1533–1592), a high prince of skepticism, was inspired by the pageant of human conflict and contradiction to think about thinking. Deriding the pretensions of human knowledge, Montaigne notes in his *Apology for Raymond Sebond* that "the wisest man there ever was [that is, Socrates], when he was asked what he knew, answered that he knew this: that he knew nothing." Elsewhere in the same work: "[The philosophers] do not want to profess expressly the ignorance and imbecility of human reason . . . but they make it plain enough to us under the appearance of troubled and inconstant knowledge." Montaigne points to "that vast and troubled sea of medical errors" as proof of the vanity and impotence of our intellect. The authorities of anatomy disagree among themselves, quackery abounds, and still "medicine is received like geometry." Our very bodies, our physical selves, refute the human pretension to knowledge.

Montaigne is the inventor of the essay as we know it—the term itself suggesting an assay, an attempt or trial, an exercise of uncertainty—and his great essay "On Experience" once again singles out physicians as pretenders.

> I do not go in much for consultations over such deterioration as I feel: once those medical fellows have you at their mercy they boss you about: they batter your ears with their prognostics. Once, taking advantage of me when I was weak and ill, they abused me with their dogmas and their masterly frowns, threatening me with suffering and then with imminent death.

The physicians with their false prognoses and shows of knowledge lack respect for Nature, which "keeps her processes absolutely unknown." Other thinkers would actually build upon doubt, as if its sands provided a paradoxically more secure foundation for knowledge than received certainties.

The founder of modern philosophy, Descartes (1596–1650), begins his project of reconstructing knowledge—which presumes, of course, that existing knowledge is a shambles and a deception—by invalidating all conclusions that do not strike the mind as indisputably self-evident, like a finding of mathematics. In giving an account of his method in the *Discourse on Method*, Descartes finds himself explaining in detail the anatomy and workings of the human heart. As, in the writings of Montaigne, the human body itself confounds the pretenders to knowledge, so in Descartes the reconstruction of knowledge requires a virtual dissection of the body. In a treatise that begins by putting into question entire categories of learning, Descartes rates the discovery of the circulation of the blood by "an English doctor"—Harvey—as glorious. The *Discourse* ends with the author confiding to us "that I have resolved to employ as much of my life as remains wholly in trying to acquire some knowledge of nature, of such a sort that we may derive rules of medicine more certain than those which we have had up to the present." Medicine is the proving ground of human knowledge. At some point the renovation of knowledge that begins in radical doubt demands a rethinking of what lies closest to home, what indeed is our home, our physical selves.

The immeasurably influential John Locke (1632–1704) administers a healthy dose of epistemological modesty in the *Essay Concerning Human Understanding*. Seeking to check presumption and, as he says, to "cure" both cynicism and sloth, Locke reminds us of the boundaries within which the human intellect necessarily does its work. There comes a time when it is best

> to sit down in a quiet ignorance of those things which, upon examination, are found to be beyond the reach of our capacity. We should not then perhaps be so forward, out of an affectation of an universal knowledge, to . . . perplex ourselves and others with things to which our understandings are not suited. . . . Our business here is not to know all things, but those which concern our conduct.

As the last sentence implies, Locke emphasizes the limits placed on human understanding not to inspire defeatism and despair but to mark that field in which we can profitably work. Accordingly, later in the *Essay* he observes that our knowledge, limited though it is, "may be carried much further than it hitherto has been" if inquirers would only commit themselves sincerely to the labor of discovery. In effect, Locke takes a middle path between the presumption of knowing all things and the idleness of not caring to know anything. "I think not only that it becomes the modesty of philosophy not to pronounce magisterially, where we want that evidence that can produce knowledge, but also that it is of use to us to discern how far our knowledge does reach"—a statement that could serve as the epigraph to many of the finasteride papers, concerned as they are to bring forth knowledge from uncertainty.

As a physician Locke seems to have been guided by the precept of avoiding harm. He recommended simple treatments, "which was quite contrary to the current practice of prescribing large doses of complicated, and often dangerous drugs," and "opposed the common, and dangerous, practice of heroic purges." "His treatment was always simple and safe." Significantly, the principles of his medical philosophy so closely resemble those of his philosophy per se that at first glance it is impossible to tell which has been modeled on which:

> General Theories . . . are for the most part but a sort of waking Dreams, with which, when Men have warm'd their own Heads, they pass into unquestionable Truths, and then the ignorant World must be set right by them: Though this be . . . beginning at the Wrong End, when Men lay the Foundation in their own Fancies, and then endeavour to suit the Phoenomena of Diseases, and the Cure of them to those Fancies. I wonder, that after the Pattern Dr. Sydenham [an accomplished clinician] has set them of a better Way, Men should return again to that Romance Way of Physick. But I see it is easier and more natural for Men to build Castles in the Air of their own, than to survey well those that are to be found standing. . . . I fear the Galenists four Humours, or the Chymysts Sal, Sulphur and Mercury, or the late

prevailing Invention of Acid and Alcali, or whatever hereafter shall be substituted to these with new Applause, will upon Examination be found to be but so many learned empty Sounds, with no precise determinate Signification. What we know of the Works of Nature, especially in the Constitution of Health, and the Operation of our own Bodies, is only by the sensible Effects, but not by any Certainty we can have of the Tools she [Nature] uses, or the Way she works by. So that there is nothing left for a Physician to do, but observe well . . .

The grounding of knowledge in observation and experience; the critique of dogmatism and hollow rhetoric; the tone of reduced ambition—these are the very marks of Locke's *Essay Concerning Human Understanding*. As a philosopher, Locke believes that ideas grow from experience; as a physician, that "whether rhubarb will purge or quinquina cure an ague can be known only by experience." With Locke and Sydenham, medicine found the modern path. Persuaded that diseases were specific entities, Sydenham sought remedies that would themselves work specifically, as cinchona (Locke's quinquina), or Peruvian bark, "arguably the first effective specific drug," was used against malaria. In the preface to his *Essay*, Locke likens himself to a mere "under-labourer . . . clearing the ground a little, and removing some of the rubbish that lies in the way to knowledge." Some decades later James Lind, having discovered that citrus fruits prevent scurvy, used the same figure in the preface to his own *Treatise on Scurvy* (1753). "Before this subject could be set in a clear and proper light, it was necessary to remove a great deal of rubbish." Convinced that ignorance of scurvy had "dangerous and fatal consequences," Lind tested his citrus-fruit theory in experiments from which today's clinical trials descend. Lind was perhaps the first to advocate the use of a placebo for control purposes in a clinical trial.

Generations of American students have traced the influence of Locke in the thought of Jefferson. Less noticed is the similarity in their medical ideas. In the spirit of Locke, Jefferson was acutely aware of the limits of medical knowledge and correspondingly cautious of aggressive medical measures:

> To an unknown disease [he wrote in a letter], there cannot be a known remedy. Here, then, the judicious, the moral, the humane physician should stop. Having been so often a witness to the salutary efforts which nature makes to re-establish the disordered functions, he should rather trust to their action, than hazard . . . a greater derangement of the system by conjectural experiments on a machine so complicated and so unknown as the human body, and a subject so sacred as a human life.

As with scurvy, though, appreciable medical progress was made during the Enlightenment. If "it is safe to speculate that in the eighteenth century a sick man who did not consult a physician had a better chance of surviving than one who did" (as Jefferson might well have agreed), this is because medicine was poorly grounded in clinical investigation; those who, following Locke, urged less theory and more observation and experiment, sought to transform medicine in effect from a harmful to a careful practice. From Locke's mentor Sydenham, "the English Hippocrates" (who believed that "the remote causes of disease were . . . beyond the range of human understanding" and approached the use of medicines cautiously), to his Enlightenment disciples ran a channel that conveyed to modernity the ancient principle of not doing harm.

The results were soon to appear. By the end of the eighteenth century many would have said that medicine itself had given people a better chance of surviving. Reviewing such remarkable advances over the course of the century as the control of epidemic diseases and the marked decline of the infant mortality rate at the British Lying-in Hospital, Peter Gay credits the gains above all to the displacement of doctrinaire knowledge by the spirit of inquiry. The knowledge of the limits of knowledge enters deeply into the tradition of modern thought and medical practice—both. The first chapter of Worthington Hooker's brilliant exposé of medical abuses in mid-nineteenth-century America concerns "Uncertainty in Medicine." Oliver Wendell Holmes taught medical students at Harvard, "The best part of our knowledge is that which teaches us where knowledge leaves off and ignorance begins." Two years before his invention of antisepsis in 1865, in a paper on the coagulation of the blood presented to the Royal Society of London, Lister inquired into the forces that prevented clotting in living tissue—"forces which I suspect will never be fully comprehended by man in the present state of his existence, and the study of which should always be approached with humility and reverence."

So constructive is the role of ignorance in the pursuit of knowledge that it has been said that

> should everything be known about a given area of science, all *scientific* activity in that area would cease, even though work might continue on the practical applications of that knowledge. Therefore, where there is scientific activity, there is partial ignorance— the ignorance that exists as a precondition for scientific progress.

Significantly, this observation appears in a thoughtful essay on medical error (as it happens, one that surmises that "only recently" did medicine turn from, on balance, a harmful to a helpful practice). The story of scientific progress which gives us an abounding confidence in medicine and inspires the expectation of one breakthrough after another—this story is also a cautionary tale warning of "the possibility of error" and the consequences of error, all the more because the individuality of human bodies, each with its own history and its own biological identity, makes bodies finally impossible to *know* through and through. (The individuality of the body also interferes with the translation of findings at the population level to any specific person.) Reflecting on the many reasons for the "poverty" of medical knowledge, another commentator, no less than the editor of the *British Medical Journal*, once cited our own "biological variability" and "the probabilistic nature of most [medical] outcomes." His intent was not to exonerate ignorance but to chastise medical pretensions of knowledge where little knowledge exists, and to remind the profession of the hazards of treating patients on the strength of poor scientific evidence. "Doctors want to believe that they know more than they do both because it feels good and because 'knowledge is power'; and the public likes the idea that doctors will cure them or keep them from death." But (as the author emphasizes) good practice demands first of all the "honest admission of ignorance." In this respect the finasteride question serves as a distillation of medicine itself. If not for the restraining effect of the sort of intellectual honesty and modesty the commentator has in mind— qualities that inspired the very tradition of empirical inquiry— doctors might now be recommending the general use of finasteride despite the little that is really known about it.

When the results of the PCPT were published in 2003, CNN asked the medical director of the Urological Sciences Research Foundation, Dr. Leonard Marks, for his comments. "I think this is a very big story. I think this is a very big day for men's health," Dr. Marks said. "The drug appears to work, 25 percent risk reduction in prostate cancer over the seven years of this study. This is not a commercial study." Asked by the interviewer about evidence of a "more aggressive form of cancer in some of the men taking" the drug in question, Dr. Marks played down that disquieting result. "I think that last point is a relatively small part of the whole story," he said. "And the reason for that is not clear yet. The numbers are small." Forming, as it were, in Dr. Marks's words is the position that the risks of finasteride are negligible because they are statistically overshadowed by the drug's benefits. The finasteride story is "very big," the numbers of aggressive cancers "small." If this view became the medical consensus (and it has its partisans in the medical literature), finasteride would be used preventively regardless of the possibility that it fuels aggressive cancer. The precept Do No Harm would be disregarded. But even as he minimized the significance of the PCPT's bad news, making it appear a mere footnote to a "very big story," Dr. Marks recognized that the reason for the elevated numbers of aggressive tumors in the finasteride arm of the PCPT "is not clear yet," and this epistemological modesty seems to have kept him from endorsing finasteride. For all his elation over the PCPT, he went on to recommend not that the drug be used generally but that its use be considered by men at high risk.

Strictly speaking, there is no reason why someone convinced the PCPT numbers are already compelling should await discoveries that might change them in the drug's favor, and yet Dr. Marks was reluctant to go as far as his enthusiasm for finasteride would take him. It is as if the deep-rooted principle that "we don't really know as much as we think we do" put this physician in touch with the also well-established precept, Do No Harm. (In the case of Locke—and few thinkers have been more influential than Locke—the principle and the precept do seem to be fruit of the same tree.) In the debate between caution and hope, it may have been an awareness of the limits on current medical knowledge that tipped the balance in favor of caution, in this case as in the finasteride papers generally. A physician who recommends a risky drug only to patients at high risk is careful of harm.

2. Informed Consent

The medical principle that at some point a preventive measure may be too harmful to employ in a healthy population is well established. By the time smallpox had been all but eradicated in the United States, the risks of vaccination itself so far exceeded the risk of the disease that it no longer made sense to vaccinate. The use of tamoxifen, for its part, is considered "safer," that is, more justifiable, in women with breast cancer because, ironically, their health is already endangered. For this reason, plans to test tamoxifen as a chemopreventive in healthy women in the BCPT met opposition. To compound matters, in many of the centers where the BCPT was conducted, consent forms did not accurately disclose the risk of blood clots associated with tamoxifen—numbers known at the inception of the BCPT. In the case of finasteride, the numerical findings of the PCPT itself are so many imponderables. If knowledge rises from experience, we have simply not had enough experience with finasteride to know much about its risks. At the time of the PCPT, finasteride after all had nothing like tamoxifen's history of use.

If a physician recommends a drug, it remains the patient's decision to take it. The sexual side effects of finasteride will be enough to dissuade some men. But a physician might not recommend finasteride in the first place if only because, given the fog of uncertainty surrounding the results of the PCPT, there is not enough good information to enable informed consent. Researchers knew that potentially dangerous blood clots occur in tamoxifen users at a rate of 1.5 percent, and the participants in the BCPT were surely entitled to that information. But exactly what the elevated rate of high-grade tumors on the finasteride side of the PCPT signifies, no one knows. In the present state of knowledge, informed consent to take finasteride is practically an oxymoron.

Moreover it is unlikely that a busy physician would do all the necessary explaining to a patient considering finasteride. Although PSA testing is by now routine in the United States, even in this case few patients receive the information necessary for meaningfully informed consent, information that at the minimum should include

(1) the possibility of both false-positive and false-negative tests;

(2) the fact that it is not currently known whether PSA screening

reduces population mortality; (3) that PSA screening may cause anxiety; (4) that PSA can detect tumors sooner than DRE; (5) that advanced prostate cancer is considered incurable; and (6) that PSA testing is controversial.

If physicians do not provide their patients with such elementary information in the case of a blood test, will they make it a practice to walk patients through the thickets of ambiguity surrounding finasteride?

Consider again the sort of uncertainty with which the entire finasteride literature is qualified. "Notwithstanding the argument that some high-grade tumors detected while the patient received finasteride may be artifactual, there are some plausible biologic hypotheses that suggest the effect could be real." "Some agents can be preventive and carcinogenic in the same organ. . . . [Such] problems highlight the profound complexity of carcinogenesis." Given the uncertainties regarding finasteride and its long-term effects—given how little, beyond its ability to shrink the prostate and its statistical association with certain outcomes, researchers themselves appear to know about its workings—would I, a layman, be in much of a position to make an *informed* decision to take five milligrams of this potent substance every day for, say, twenty years? At best the statistics associated with the finasteride regimen could be explained to me. But if a mixed lot of numbers were enough to yield a good judgment in such a case, the literature on finasteride wouldn't be as divided as it is. The finasteride question wouldn't exist. It bears remembering that the principle and case law of informed consent evolved in the context of treatment, not the more speculative one of prevention, and dictated that patients undergoing a particular procedure be advised of its known risks. In each of two landmark legal cases, patients had not been advised of a risk of paralysis known to attend a procedure. That so much about finasteride's effects is unknown or in dispute, in itself argues against using the drug, or at least using it generally and speculatively.

If a physician who took informed consent seriously might hesitate to recommend finasteride, a drug whose effects are poorly understood and whose results are a tale of ambiguities, for purposes of prevention, what of the use of the drug to reduce enlargement of the prostate? It was the drug's success in that department that led to its being tested as a chemopreventive agent in the first place. Ironically, with the PCPT it emerged that men who had been taking finasteride for BPH stood exposed to a risk of high-grade cancer unsuspected when the drug was approved. These men, having already taken finasteride for some years under medical monitoring, have at least the information of their own experience to guide them as they decide whether or not to continue. (Other treatments for BPH are now available.) Because men taking five milligrams of finasteride to shrink the prostate face the same risks as men taking same drug at the same dosage purely for prevention—a point the press did not fully appreciate with the report of the results of the PCPT in 2003—physicians may hesitate to prescribe finasteride for new cases of BPH while keeping a close eye on those patients already on it.

3. ONLY AS EXPECTED

For a doctor weighing whether or not to prescribe, or a patient whether or not to take, finasteride, the question of how much is known about the drug and its effects would be material. Considering how many competing explanations of the PCPT data are in play in the medical literature, and how steeped in uncertainty the entire literature is, it is ironic that retrospective analyses of the PCPT sometimes imply that things in the study worked out exactly as predicted. "If one assumes that low grade cancers are effectively prevented and that high grade cancers are not, one would expect, based on a [detection bias], a higher proportion and number of high grade cancers in the patients treated with finasteride. . . . This, in fact, was observed." But was this expectation something researchers actually carried with them into the PCPT, or something invented after the fact to make sense of the data? Surely the latter, for who would design a massive, costly, laborious study expecting such a perverse result as a reduction of the less dangerous, hand in hand with an increase in the more dangerous, cancers? Like bad historical writing that makes events seem inevitable in retrospect, the fiction of a confirmed expectation makes the PCPT appear less disturbing and enigmatic than it actually is. Statements of the same kind, implying that the results of the PCPT were only to be expected, appear here and there throughout the finasteride literature. But if the results of the PCPT were predictable, why were they not predicted, and why are they so disquieting and subject to such conflicting interpretations?

A paper reviewing the PCPT, and predicting results that occurred, argues that "in a study demonstrating a lower prevalence of prostate cancer in finasteride vs placebo treated men one would expect there to be more suspicious PSA increases in the placebo group. This was indeed the case in the PCPT." A disturbing enigma begins to seem like a foreseeable outcome. Another paper, this one examining the PCPT data three years after the end of the study, concludes that increased sensitivity of PSA in the finasteride group introduced a bias that "would be expected to contribute to greater detection of all grades of prostate cancer with finasteride," almost as if this result had been foreseen in detail. Clearly, though, the expectation of a distortion in the data arose after the fact; the phrase "would be expected to contribute" amounts to a rhetorical way of saying "probably contributed." A press report on the latter paper builds up that much more of an illusion that the PCPT merely confirmed expectations: "Because of the increased sensitivity of PSA testing with the use of Proscar, the increase in aggressive tumors could only have been expected," wrote Ian Thompson, M.D., of the University of Texas, and colleagues, in the August 16 issue of the *Journal of the National Cancer Institute*. Not only is a worrying event—the increase in aggressive tumors, the result that, more than any other, has ruled out the general use of finasteride—reduced to the ordinary, but an unforeseen event is made to seem entirely predictable. Rhetorically, the "could only have been expected" construction reads like damage control. The results of the PCPT were so far from being what could only have been expected that it is hard to imagine the study would ever have been conducted if its outcome had been foreseen from the beginning.

Even knowing that autopsy reveals unsuspected prostate cancer, who would have predicted that biopsy would yield so high a rate in the PCPT control group? Who would have supposed that a finding so filled with disturbing import would emerge on the placebo side of a study at all? The designers of the trial anticipated that over its seven years, 6 percent of the placebo group would be diagnosed with prostate cancer. The final figure turned out to be four times that. It is as if the vertical increase in detected cancers that accompanied the advent of PSA testing, an increase so dramatic that it is "unsurpassed by any other tumor in the history of modern public health statistics," were returning at an even higher level. In part it is because PSA testing brought about an unforeseen epidemic of diagnoses that the prospect of a way out of that epidemic, via finasteride, is so appealing. The equivocal results of the finasteride trial were themselves unforeseen, however.

The pretense of the predictable in the finasteride papers is all the more ironic in view of the course of events that led to the synthesis and marketing of finasteride in the first place—events that seem predictable only after the fact, if then. The story of the potent pill taken by half the men in the PCPT began somewhere in the villages of the Dominican Republic. "No one could have anticipated the revolution in prostate treatments that followed the discovery of pseudohermaphrodites in the Dominican Republic more than 30 years ago." The finasteride papers may argue that this or that finding of the PCPT is only to be expected, but the finasteride story itself, dating as it does to such an unlikely point of origin, tells of the unexpected. Reviewing what was known about those born with ambiguous genitalia as a result of a 5 alpha-reductase deficiency, the author of a 1994 paper in the *New England Journal of Medicine* concluded, "One would predict, therefore, that pharmacologic inhibition of 5 α-reductase would prevent the development of prostate disease, male-pattern baldness, and the growth of facial and body hair but would not affect libido, potency, or male musculature." In the event, finasteride does depress libido and impair potency in appreciable numbers of men, a consideration that will surely weigh with many.

Ironically, too, amidst all the retrospective prediction in the finasteride papers little notice is taken of those who actually did predict that the drug could induce cancer, such as Maarten C. Bosland, who wrote in 1992, "Some of the most promising agents for chemoprevention of prostatic cancer, *i.e.*, 5α-reductase inhibitors and retinoids [compounds related to vitamin A], may enhance rather than slow down the progression of prostatic cancer." But who at the time supposed that finasteride would do both? To my knowledge, no one foresaw this Janus-like outcome. In an editorial in the same issue of the *New England Journal of Medicine* that carried the results of the PCPT, it was duly noted that the 25 percent reduction in the incidence of prostate cancer was "contrary to the expectations of many experts." No less unexpected were the more disturbing findings of the PCPT. Two years before the results of the PCPT were reported, a paper on prostate cancer prevention in the journal *Urology* confidently affirmed the safety of finasteride.

> In data derived from many thousand patient-years of finasteride treatment in double-blind trials and millions of patient-years experience obtained from postmarketing toxicity surveillance, no serious drug-related safety concerns have arisen.

What all that experience with finasteride did reveal was higher rates of impotence and other sexual side effects, which in the eyes of medicine may be sorrowful life-style impairments but not medically serious like a question of safety.

In the case of the PCPT's sister study, which was conducted in spite of projections of significant harm, "when the study began, proponents suggested that [tamoxifen] may reduce the incidence of breast cancer in [high-risk] women by as much as 30 to 40 percent, while critics claimed that this figure was too high." The actual figure turned out to be 49 percent. As for the adverse effects of tamoxifen in the BCPT, the *New York Times* reported the directors of the National Institutes of Health and the National Cancer Institute as claiming that they "did not exceed expectations," as if that made them less troubling. Nothing new here.

Sometimes the results of a clinical trial not only exceed but defy expectations. The Vioxx Gastrointestinal Outcomes Research (VIGOR) study, published in 2000, did not include good information about the incidence of heart attacks because, as its name suggests, it was not looking for heart attacks. "Such events had not been prospectively defined." Four years later Vioxx was withdrawn from the market when analysis of data from already completed studies established that the drug was in fact associated with a significantly higher risk of heart attacks. Unlike the VIGOR study, the retrospective analyses knew exactly what to look for.

What do things look like and feel like before they are discovered to be such as "could only have been expected"? Some answers are found in the literary form whose name recalls the word "new," the one form that makes a specialty of unfolding knowledge gradually and subjecting readers to the same uncertainties and errors as those they are reading about: the novel. The modern novel is proverbially dated to the time of Locke and characteristically takes the same attitude toward romance as he does toward "the Romance Way of Physick" and "Castles in the Air." The novel might also be considered Lockean in its strong interest in the nature, limits, and fallacies of human knowledge. Great novels have been written as if with the purpose of humbling our pretenses of knowledge and confounding the illusion that events unfold in ways that could only have been expected. In these novels many expectations are disconfirmed, many theories dashed by experience.

> A character who believes that love is purely physiological winds up falling desperately in romantic love, as in *Fathers and Sons*; a hero who believes that all is permitted comes to feel guilty for a crime in which he is, at most, an unwitting accomplice, as in *The Brothers Karamazov*. Whatever experience has led to a given theory, other, surprising experiences reveal its shortcomings. In this way purported Truth becomes fallible opinion; there are no final truths.

Some realist novels make a principle of epistemological modesty, virtually dramatizing our liability to error, making us revise what we thought we knew, and reminding us that things in real time do not in the least possess the order and "inevitability" they acquire in retrospect. Despite the legendary chasm separating the sciences and the humanities, the novelist no less than the scientist embraces the principle that "we really don't know as much as we think we do." Sydenham himself, Locke's medical mentor, when asked what medical books to consult, reportedly answered, *Don Quixote*—the work that presides over the tradition of the modern novel—his point being that just as Don Quixote was led astray by books, so the student of medicine ought to put away his books and attend to the lessons of experience.

The scientist who proclaims an enigmatic result "just as expected" might learn from works of fiction in which very little happens as expected. In the hands of a novelist really committed to the principle that things rarely occur predictably, such a novelist as Tolstoy, even the past comes to seem unexpected. Trying to make sense of the historic PCPT with its mixed data and messages, we are like readers caught up in a novel of conflicting possibilities—except that the ending of *this* tale hasn't been written yet. Surrounded by bewildering ambiguity, finasteride offers a particularly vivid instance of the general medical principle that "the effect of a given therapeutic intervention on a given patient is always to some extent uncertain." The need to proceed with due caution, that is, with regard for the Hippocratic principle of not doing harm, follows irresistibly.

It is only fitting that of Hippocrates himself, the reputed founder of medicine, virtually nothing is known.

[7]

Prevention Gone Too Far

❦❦❦❦

IF PREVENTIVE MEDICINE grew from practices such as quarantine designed to control the spread of pestilence, ideal communities in the tradition of More's Utopia resemble quarantined societies, set apart from the world at large and somehow secured against its corruption. In the case of Utopia the infectious evil guarded against is, above all, greed. "So long as private property remains, there is no hope at all of effecting a cure and restoring [European] society to good health," and the Utopians do away with private property and the parade and poverty that go with it. Not content to abolish all this by decree, the Utopians make sure to eradicate the very desire for wealth in every generation. It is as if they abolished property continually. Lust for wealth being the root of evil, the Utopians take particular care "that no one shall value gold and silver" by conditioning citizens from infancy to despise all such objects of greed. Children's acquired distaste for the metal is described in unusual detail by More, exemplifying as it does the preventive logic of the Utopian system. In order to drive home the idea that a liking for gold is something one outgrows at an early age, Utopian pacifiers are made of that precious metal, and over time the young get used to seeing gold on slaves and criminals as well as chamber pots. With gold transformed into a kind of fetish of disgrace, the very possibility of such greed as infects and ruins Europe is uprooted in Utopia. Having expelled greed and "torn up the seeds of ambition and faction at home, along with most other vices," the Utopians "are in no danger from internal strife, which alone has been the ruin" of many nations. The very stability of the Utopian commonwealth testifies, then, to the efficacy of a sweeping regimen of prevention.

Among the heirs of the utopian tradition are those who demand a shift toward preventive health measures, but define prevention as the rooting out of everything wrong with our way of life. In the 1990s some of this school protested tamoxifen not only as a dangerous substance passed on the public by a profiteering corporation but as a kind of political sedative—a pill designed to keep people from realizing that the causes of breast cancer lie not in biology but in the way we live and the perils we live among. "The profit potential in a patentable pill to prevent breast cancer could draw industry interest, whereas research into the environmental causes of breast cancer or the prevention potential of a low-fat diet, for example, could only tell women what to avoid, not what to buy." Fully worked out, this indictment of a toxic way of life and the massive deception it rests on would resemble Rousseau's jeremiad against civilization:

> If you consider the anguish of mind which consumes us, the violent passions which exhaust and grieve us, the excessive labours with which the poor are overburdened, and the even more dangerous laxity to which the rich abandon themselves, so that the former die of their needs while the latter of their excesses; if you think of the monstrous mixtures they eat, their pernicious seasoning, their corrupt foods and adulterated drugs; the cheating of those who sell such things and the mistakes of those who administer them . . . if you take note of the epidemic diseases engendered by the bad air where multitudes of men are gathered together, take note also of those occasioned by the delicacy of

our way of life . . . you will see how dearly nature makes us pay for the contempt we have shown for her lessons.

If this represents the world as it is, the world must be transformed down to the last detail, as in Utopia. Those who would extirpate the root causes, real or imagined, of cancer demand a transformation no less complete. That is what their rhetoric of prevention comes to.

With chemoprevention in its infancy, it may seem the wrong time to warn that dreams of prevention can go too far. But the very discovery that aroused interest in 5 alpha-reductase inhibitors in the first place—a cluster of pseudo-hermaphrodites who were free of prostate cancer because they had virtually no prostate gland—reminds us that absolute prevention may not be a thing to be wished for. Castration will prevent prostate cancer. Fully half of all cancers detected in the PCPT eluded discovery by regular digital rectal exams and PSAs for seven years, only to show up in end-of-study biopsies, yet no one suggests that men simply skip screening and get themselves biopsied annually in the name of prevention. That would be going too far. Given that "as many as 32%" of men in their thirties who die of trauma show incipient prostate cancer at autopsy, in order to suppress cancer in its early stages (which is what prevention implies), men would have to begin taking finasteride, an anti-androgen, as early as that. That too would be going too far. Overzealous prevention has its costs. Perhaps the crowning tragedy of the DES story is that so many women who really had nothing wrong with them, who were taking the drug purely speculatively—preventively—in the belief that it somehow made for a healthy pregnancy, ended up visiting misfortune on their own children.

Now that emphasis in medicine itself "is shifting increasingly toward the concerns of society," the aim of preventing disease readily broadens into larger visions, and some enthusiasts dream of establishing a comprehensive preventive regime without considering the costs of overzealousness. Convinced that many if not most of our ailments are self-inflicted, preventionists inveigh against alcohol abuse, smoking, overeating and bad eating, and lack of exercise, among other bad habits. While no one would deny that smoking leads to lung cancer, overeating to obesity, and all the rest, the composite picture of citizens who blindly pursue self-destruction as if it were the object of their life jeopardizes the very principle of self-government by depriving those persons of the last modicum of competence and reason. Nevertheless this image of one's fellow citizens courting suicide "enjoys widespread currency among academics, physicians, policymakers, and [ironically] the general public" in Britain and the United States—even though people in both countries are in fact living much longer than they did a century ago.

Surveying the theory and practice of cancer prevention, two researchers call attention to the need for "additional development" in many areas, including "medical school and pre- and postdoctoral cancer prevention training programs; youth education programs on physical activity, sun exposure, diet, and tobacco use; research funding; legislative and public health policies; [and] health insurance policies." Whether or not they realize it, the authors have in mind nothing less than the reform of human life in the name of health. If all things from sun exposure to insurance policies relate to prevention, then we might as well just say that life in general is to be brought under medical cognizance. We are to live under the eye of enlightened medical guardians, to the end of making ourselves happy and safe. The enthusiasts of prevention follow in the line of many others over the last century, from would-be social planners to medical thinkers with "with visionary leanings," all animated in one way or another by the hope of managing human existence and elevating medicine from a practice that deals with disease to one that keeps disease from coming about, from a profession with a kind of service role to a "universal" monitor.

The medicalization of life has already progressed to the point where "everyday experiences like insomnia, sadness, twitchy legs and impaired sex drive . . . become diagnoses: sleep disorder, depression, restless leg syndrome and sexual dysfunction." The next phase in the manifest destiny of medicine would be some sort of militant, society-wide campaign for health—a campaign also implied in some of the finasteride papers, for (even assuming that the risk of high-grade cancer proves false) only if prevention became a crusade would entire categories of men risk the drug's sexual effects for the sake of its anti-cancer potential. But what if the world at large were transformed into a hospital without walls, or for that matter a childproof household with cupboards locked, floors padded, stairs barred by safety gates, and doings properly supervised? Before launching any such campaign against ourselves, we should recognize that at some point it is better to accept human life as it is.

In his later years Tolstoy came to be convinced, like the Utopians, that possessions are the root of evil, so much so that he abandoned property concerns to the wife he hated (but could not tear himself from until his last days) and regarded his very writings as morally the property of humanity. Tolstoy could have signed his name in blood to the statement that "So long as private property remains, there is no hope at all of effecting a cure and restoring society to good health," and in his disillusionment with a diseased world he portrayed human life, in a powerful fable, as being so fraught with moral risk that the best preventive would be an early death. Rousseau exposed what he took to be the root causes of bad health; Tolstoy, a disciple of Rousseau, went further by suggesting that it would be better to die young—very young—than to live in a sick world. As the ideal of prevention grows ever more attractive and compelling, Tolstoy's story "The Prayer" is worth contemplating as an instance of the ideal of prevention gone terribly awry.

In "The Prayer" a mother distraught over the death of her three-year-old son Kostya is visited in a dream by a higher being who both rebukes and consoles her.

> "You poor, blind, impertinent, and conceited creature," says this being. . . . "Don't you understand that if, from nothing, your son grew to become who he was, he would not have stopped changing for a single minute and would not have remained the same as he was when he died. . . . From a child he would have become a teenager, then a young man, then an adult, than an aging man, then a very old man. You do not know what he would have become if he had lived. But I know."

And the mother is granted a vision of what her son would have become: an old man sunk in debauchery.

> And then the mother sees a small, brightly lit room in the back of a restaurant. . . . Sitting at a table that holds the remains of supper is a rather fat, wrinkled old man with his nicely cut mustache curled upwards, an unpleasant old man trying to look younger. He sits deeply in a very soft sofa, and he peers with greedy, drunken eyes at a corrupt, heavily made-up woman baring her plump white neck, and his drunken tongue shouts and jokes rudely, cheered on by the laughter of another couple of the same sort.

Preventionists today preach against fat, inactivity, and poor dietary habits. The aged Kostya is envisioned a fat man sunk in a couch in a restaurant with a certain resemblance to a house of prostitution. In time the corruptions of human life would have claimed Kostya.

Awakening, the mother glances at her dead child and is "immediately astonished by the terrible resemblance of this sweet small corpse to the face of the terrible old man she saw in her sleep. . . . And then she weeps for the first time. She cries not with hopeless tears, but with tears of joy and relief and release." The early death of her son is thus revealed as a merciful act of prevention. "The Prayer" is intended to console, to teach submission, and to justify the ways of God, but its extremism somehow undoes these pious objectives. It is one thing for a man to learn by revelation that his young child died in a state of grace, as in the late-medieval dream-poem "Pearl," another for a mother to learn by revelation that human life is vile. It is one thing for Tolstoy's own Ivan Ilych to remember his mother on his deathbed and mourn the loss of his childhood, another for a mother to foresee her child's degeneration with time and rejoice in his early escape from this world as if he were spared a wretched fate. As humane as "The Prayer" is intended to be, it is inhuman in its omniscience and misanthropic in its judgment of life. It also leaves the impression that millions of others like Kostya might beneficially have been prevented from growing old. Contrary to its own intention, what Tolstoy's fable teaches is that dreams of prevention should be qualified with common humility and an acceptance of things as they are—a moral corollary of epistemological modesty.

*

Regarding Tolstoy's fable, a point so obvious that it may go overlooked should be made. Kostya did not ask to be spared the shame and corruption of adult life. He was simply the passive recipient of a blessing.

Today, even as bioethics insists on respect for patients as autonomous persons, many still seem to think of ordinary citizens as incompetent folk whose lives need to be regulated, and decisions steered, by the hand of medical benevolence. All the more significant in this context are findings regarding breast cancer patients who take perhaps the most radical of preventive measures, that of having the opposite breast removed along with the stricken one, to preempt the possibility of "contralateral" breast cancer at a later date (a risk that runs from 40 to 65 percent in patients with a common genetic mutation). Recent studies suggest that breast cancer patients who choose this course "fare as well as those who do not in the short and long term." The key word seems to be "choose." Women who are informed of their probabilities but choose their own course, who (in the journal-jargon) take part in "active decision-making," seem to reconcile themselves to their fate. It is this element of choice that Kostya is denied and that is likely to be sacrificed in schemes for the reform of human life descending from the tradition of More's Utopia.

[8]

Breakthrough:
"The Present Is Obsolete"

◦Ο◦Ο◦Ο◦

TO THOSE who believe that tradition is a drag on human life (even if they themselves continue to draw from tradition), it seems only natural that progress should arrive by way of a sharp departure from existing practice: a breakthrough. With the cultic authority that accrued to Thomas Kuhn's *Structure of Scientific Revolutions* (1962)—the work that inspired the vogue-concept of a "paradigm shift"—the romance of the breakthrough acquired a kind of scientific cover story. Advocates of all kinds of departures could now claim that science itself, the proudest achievement of Western culture, advanced not by degrees but sporadic leaps and breaks. By pure happenstance, the advent of the notion of the paradigm shift more or less coincided with the acclamation of a new anti-cancer agent, interferon, as a wonder drug with potential to break the impasse in cancer treatment. It failed to do so, though it "led to endless hype about the latest drug 'breakthroughs'; to personality cults among researchers who suddenly found themselves on the front cover of *Newsweek*; and to the contamination of the whole field [of cancer research] by big money."

Such is the romance of the paradigm shift that some medical ethicists now tell of a great seismic change—dating, as it happens, to within a few years of Kuhn's book—in which the old medicine, arrogant and heedless of the individual patient, fell, displaced by a new medicine at last respectful of rights. This sort of story line has great simplicity and great appeal. For at least a generation now, the united power of advertisements and popular culture has celebrated the advance that shatters existing models and inaugurates a new era. One thinks of the famous Apple ad of 1984 introducing the Macintosh, with a woman hurling destruction at Big Brother: an ad that not only depicts the shattering of things established but actually explodes conventional restraints with its astonishingly brazen theft of Orwell. Testifying before the Senate Subcommittee on Appropriations on May 7, 1997, one of the directors of the PCPT reduced the Apple message to an aphorism. "The present is obsolete," he said. A variant of the breakthrough myth is the story line of pop psychology, in which a habit-bound person, trapped in a life of unhappiness and inauthenticity, finds liberation by summarily wiping the slate clean. Jeremy Bentham, as we know, wiped the slate clean, "begin[ning] all his inquiries by supposing nothing to be known on the subject"; but by now the idea of zeroing out the past has become so commonplace that breakthroughs are expected events, disappointing when they fail to occur. Medical breakthroughs in particular are looked for. Some thirty years ago a researcher ended his paper on the natural history of prostate cancer with the hope that it would one day be possible "to inject more science into the extant art of treatment of the prostatic cancer patient and substitute an era of cold fact for the present era of heated opinion," as if history advanced by seismic displacements. Since then, PSA has arrived, as has finasteride, but we remain in the era of opinion.

One of the main reasons so much hope has been invested in chemoprevention is that, ideally, the prevention of cancer would cut the all-too-familiar knot of diagnosis and treatment—simply abolish the existing paradigm. But fixed ideas and ideological fashions like a belief in breakthroughs are not in the spirit of science. At the start of the tamoxifen trial a *New York Times* writer predicted that "If the five-year study shows that the hormone prevents cancer, millions of women will probably be advised to take it for the rest of their lives," but while the study did vindicate the drug's preventive potential, it is not in general use. The expectation of a breakthrough was fallacious. Around the time it began, defenders of the tamoxifen trial pointed out that many doctors were already prescribing the drug to women at high risk of breast cancer, and given that it had already crept into use as a preventive, it was a good idea to test its effect as such under controlled conditions. Largely as a result of the double-edged findings of the BCPT, tamoxifen is now recommended as a preventive only for women at high risk of breast cancer. In other words, despite the drama of the BCPT, and despite the rhetoric of breakthrough, things at the level of practice seem to have remained much as they were before the BCPT. A powerful drug is used sparingly.

We speak too loosely of breakthroughs, as if revolution had become the conventional unit of scientific progress. Continuous dialysis, first performed in Seattle in 1960, was so important a medical advance that its implications brought into being the field of bioethics. Dialysis itself, however, had been invented in 1943. Tamoxifen was originally synthesized for purposes of birth control. PSA, the instrument of prostate cancer detection, was originally of interest as a tool in rape investigations. It was approved by the FDA in 1986 not for the purpose of screening millions but as a way of monitoring the progress of prostate cancer. Chemotherapy evolved from chemical warfare agents on the order of mustard gas. Originally acclaimed as a "miracle cure for rheumatoid arthritis," steroids proved useful in the treatment of dozens of disorders. There have even been inverted breakthroughs, such as the introduction of mood drugs in the 1950s which "opened up a whole new dimension of popular, patient-driven illness," that is, made it possible for broad categories of people to be ill and seek medical relief. Few breakthroughs can compare with penicillin, its effects so striking that they obviated the necessity of a clinical trial.

> No controls . . . are essential to prove the value of a drug such as penicillin, which quickly reveals dramatic effects in the treatment of disease. Such dramatic effects occurring on a large scale and in many hands cannot be long overlooked. Unfortunately, these undeniable producers of dramatic effects are the exception rather than the rule, even in these halcyon days of antibiotics.

So wrote A. Bradford Hill, the designer of the first randomized clinical trial and the co-discoverer of the statistically undeniable association between smoking and lung cancer, in 1951.

Finasteride, which unlike penicillin did have to be tested in a clinical trial, is one of those drugs that prove not to be producers of unequivocal effects. Instead of slicing through conceptual problems in the manner of a breakthrough, the PCPT introduced a Gordian knot of new problems. (There may be good biological reasons why success in treating BPH did not translate into a breakthrough in the prevention of prostate cancer. As a number of papers acknowledge, no connection between BPH and prostate cancer has been shown to exist. The two diseases typically originate in different parts of the prostate.) Flying like a sort of banner over one of Merck's web pages is the statement, "We look forward to the opportunity to bring the next breakthrough medicine." The page in question advertises Proscar as a treatment for BPH while carefully avoiding any implication that the same drug at the same dosage reduces prostate cancer. It was surely because the PCPT did not provide an unqualified breakthrough that the media quickly lost interest in finasteride—though the fact that the data did not conform to the story line of our expectations may have been the real story.

Considering that there is no absolute line dividing what is cancer from what isn't, it is perhaps no wonder that the model of the clean break, the definitive advance, the categorical shift, the solved problem does not apply very well to the existing state of breast and prostate cancer knowledge. I have heard that abnormalities classified as prostate cancer today would not have been so labeled a generation ago. The very criteria of cancer seem to move with the times, so that "many men who would have qualified as controls in previous genetic and epidemiological studies are now known to have prostate cancer as a result of PSA screening." Cancer can elude visual definition too, as when cells under the pathologist's microscope cannot be identified positively as either cancerous or noncancerous—not a rare event. (During the PCPT, "discordant interpretations were arbitrated by a referee pathologist.") Because cancer is apparently of gradual onset, even its biological definition is blurry, with some, for example, questioning whether DCIS really amounts to breast cancer. "The molecular hallmarks of cancer development . . . are frequently present in both cancer and precancer," the "treatment" of precancer being tantamount to prevention. It appears that finasteride may, accordingly, reduce prostate cancer by suppressing its precursor, high-grade prostatic intraepithelial neoplasia (HGPIN), "a preinvasive intermediate step in prostatic carcinogenesis. . . . Finasteride appears to have an effect across a broad spectrum of neoplastic [that is, abnormal] stages ranging from premalignant to invasive disease."

In a 2002 review of the theory and practice of cancer prevention, the two authors (one of whom was also a co-author of the *New England Journal of Medicine* paper on the PCPT) note that

> Molecular targeting research has brought about a revolution in drug development and is blurring the distinction between malignancy and premalignancy and between cancer therapy and pre-

vention. Tamoxifen was developed first for cancer treatment and later for prevention. Indeed, tamoxifen almost certainly both treated and prevented microscopic, subclinical disease in the BCPT, illustrating that therapy and prevention are also blurring at the clinical level.

A revolution is proclaimed but an evolution described, with tamoxifen being used first for treatment and later for prevention. If at some level treatment and prevention come to the same thing, this blurring of a distinction (corresponding to the blurry distinction between malignancy and premalignancy) suggests the very opposite of a clean, definitive break with existing theory and practice. Just as tamoxifen's use in the treatment of breast cancer suggested its potential for chemoprevention (for it was found to halve the expected incidence of cancer in the opposite breast), it was finasteride's proven ability to reduce the enlarged prostate that led to its being tested as a preventive agent. To the many parallels between the sibling diseases, prostate and breast cancer, we may now add this one: that in each case a drug tested for prevention was already in use, in defiance of the breakthrough model which dictates that advances should be "revolutionary." In the case of both diseases, too, the advent of aggressive screening failed to bring about a breakthrough, if a breakthrough means in this context, first and last, a dramatic reduction in deaths.

As noted earlier, another drug (or in this case family of drugs) with chemopreventive potential is also already in use for another purpose. Statins were found to be associated with a lower incidence of cancer, and impressive reductions of some particular cancers, as a result of clinical trials designed to test their efficacy against cardiovascular disease, not cancer. Their showing against cancer came solely as an unintended consequence. Together with the evolution from treatment to prevention itself, statins suggest that progress against cancer may lie not so much in dramatic breakthroughs, however romantic, as in incremental advances. After noting that clinical trials have identified "new uses for older drugs," a physician writing in 1998 pointed out that

> Advances in cancer treatment generally come in steps and slowly. It would be astonishing if any single therapy worked against the scores of diseases known collectively as cancer. So far, anticancer drugs generally have proved more effective in combinations than singly.

Prevention, in turn, may advance into a sort of monitored holding operation, with anti-cancer drugs delaying the growth of tumors "much as insulin keeps diabetics alive without curing diabetes." This much, however, is clear: in weighing the finasteride question with all of its ambiguities (and similar controversies are likely to follow), it is better to be mindful of the traditional distinction between the certain and the uncertain than to ignore it as if it belonged to some imprisoning conceptual scheme lately broken through.

Announcing the results of the PCPT, a CNN reporter declared, "More than 220,000 men will be diagnosed with prostate cancer this year. And the American Cancer Society says almost 30,000 of them will die. There's no sure way to prevent it until now." Enthusiast of an imaginary breakthrough, the correspondent states categorically that finasteride prevents prostate cancer. In point of fact, as a physician from the National Cancer Institute duly noted in the CNN report, finasteride reduced the incidence of prostate cancer in the PCPT by a quarter; it did not eliminate it outright. Only someone looking for a quantum leap in cancer prevention—someone captivated by the romance of the breakthrough—would misinterpret the PCPT as a kind of total victory and finasteride as a "sure way to prevent" prostate cancer.

Some would say that modern medicine itself came about as a result of a breakthrough centuries ago—the overthrow of the authority of Galen in the name of the empirical investigation of nature. Among the innovators was the sixteenth-century physician Paracelsus, who despite his renown as a quack conducted valuable trial-and-error experiments and arrived at some solid findings. In a gesture comparable to Luther's burning of a papal bull, Paracelsus burned texts of Galen and Avicenna; but the distinction between textual authority and empirical observation was not in and of itself a breakthrough. Over a century before it had been cited explicitly, and sensationally, by Chaucer's Wife of Bath as the source of her right to report her own trials, errors, and personal heresies to the world. The Wife lays her claim to knowledge of marriage and its griefs not on the basis of the written word, whatever its authority, but on the strength of her own abundant experience, her firsthand observation. "Experience," she says, "even if there were no books in the world, would be quite enough for me to speak on the woes of marriage," and she proceeds to wage war on ancient authority and the written word, reporting in the course of her wayward narrative that she made her fifth husband, a cleric, burn his most precious book. There is textual precedent for the sixteenth-century revolt against textual knowledge. Paracelsus himself seems to have been a somewhat medieval character, his chemistry mixed up with alchemy. It is possible to overdo the story of the Renaissance breakthrough. Columbus was inspired by the travel narrative of Chaucer's contemporary, the shadowy John Mandeville. Nevertheless, as exaggerated as it is, the story line of the Renaissance break with the past has great appeal even now, perhaps because it serves to prefigure and ennoble our own revolt against the past. But the past continues to instruct us.

As postmodern and at times surreal as it is, the finasteride question still has us deliberating in common on a matter of uncertainty and dispute, as has been done since antiquity. Says Aristotle, "The duty of rhetoric [or argument] is to deal with such matters as we deliberate upon without arts or systems to guide us. . . . The subjects of our deliberations are such as seem to present us with alternative possibilities"—as the finasteride debate presents the options of using the drug generally, using it selectively, or using it not at all. "For it is about our actions that we deliberate and inquire, and all our actions have a contingent character," meaning they could be otherwise. Fittingly, therefore, much of rhetoric concerns probabilities, or things that happen "for the most part." The entire finasteride debate is probabilistic, which is not to say that anyone trying to make sense of the finasteride question had better consult Aristotle. But if we disregard the sorts of elementary distinctions drawn by Aristotle and etched into our tradition and literature, we will be left in much the same state as the CNN reporter, apt to confuse a partial and unclear result for a definitive one, and befooled by our own investment in the fantasy of medical revolution. While the authors of the finasteride literature never cite tradition, the care they take in weighing evidence and framing conclusions did not come from nowhere. "Modern medicine inherits a long tradition in Western culture. Even today, when younger members of the profession know almost nothing of that history and when it seems fashionable to pretend that there was no history before antibiotics or before microbiology, the long tradition shapes medicine." In the finasteride debate this tradition speaks, cautioning us against overstating our knowledge and reminding us of everything at stake in our conclusions.

[9]

"I Will Abstain from Harming Any Man"

∘❍∘❍∘

1. HEALTH AND HARM

"The operation was a success but the patient died." The sly old joke comments on the physician's pride in his technical skill and imaginary victories, though it also glances back to the time when medicine had little to offer but illusions, and people healed despite the medical art and not because of it. Over the centuries when patients were bled and purged as if some archaic rite of purification were being enacted on their own bodies, the centuries before the discovery of germs, before antisepsis, before anesthesia, many indeed must have died at the hands of medicine. And the satirist, for one, did not let the doctors forget it. Among the rational horses Gulliver finds "neither physician to destroy my body, nor lawyer to ruin my fortune." Of physicians Mosca in Ben Jonson's *Volpone* says, "they flay a man / Before they kill him." When his interlocutor comments, "It is true, they kill / With as much licence as a judge," Mosca disagrees. "Nay more; / For he but kills, sir, where the law condemns, / And these [physicians] can kill him too."

The Hippocratic tradition seems to have been much concerned with keeping the physician above suspicion, permitting the use of anodynes, for example, "only if they could be rendered completely non-poisonous." But while the Hippocratic Oath pledges the initiate neither to "give a fatal draught to anyone if I am asked" nor to suggest the same—the clause that follows the vow not to harm—the distinction between a medicine and a poison can be delicate. In antiquity, "the paradox of the murderous healer was the subject matter of a great many rhetorical exercises." One of the stories nested in the *Thousand and One Nights* tells of a sagacious physician who cures a king's leprosy by means of a medicinal polo mallet, only to avenge his own execution by his ungrateful client by getting the king to consume a lethal poison. The very symbol of medicine, a snake coiled around staff, is suggestive of deadly poison. A dangerous power is placed in the physician's hands. At around the same time the principle of avoiding harm was becoming established in medical practice, the use of poisonous substances as medicines came under scrutiny and criticism, as when Oliver Wendell Holmes reminded medical students at Harvard that "the most intelligent men in the profession have gradually got out of the habit of prescribing . . . powerful alien substances [like mercury, lead, arsenic, and antimony] in the old routine way. . . . Still, the presumption in favor of poisoning out every spontaneous reaction of outraged nature is not extinct in those who are entrusted with the lives of their fellow-citizens." Some abuse that trust, an extreme case being that of Harold Shipman, the English physician who drugged to death untold dozens of elderly patients under the guise of treating them.

Although physicians have been feared, distrusted, and satirized, this is not to say that the world gave up on the ideal of a physician who did not harm. In the Middle Ages, when the Greek art of medicine, modified by Arabic and Jewish learning, entered Christian Europe, it was possible to refer to Christ as the physician of souls. The tradition of literature for its part lends rhetorical support to the Hippocratic ideal insofar as it delegates cutting to barbers and makes apothecaries and charlatans, not physicians, dispensers of poison. So too, in portraying physicians as destroyers of health, or indeed pretenders who accomplish nothing, literature appeals implicitly to the ideal of the physician who might actually comfort, sustain, or heal.

In one of the *Canterbury Tales*, a powerful story with sources reaching back into the world of oriental tales and a descendant in modern cinema, three young revelers in a time of plague discover a trove of gold, whereupon each begins to plot against the others in order to keep it for himself. The youngest, dispatched to town to get provisions, heads straight to a "pothecarie"—not a physician—and buys poison, which he mixes with wine. Upon his return he is promptly slain, his killers celebrating the job by drinking the wine: and so with the unwitting assistance of the local pharmacist, the three destroy one another. For his part, the teller of this tale, the Pardoner, professes to own a bone that once belonged to a holy Jew and which, when dipped in a well, changes the water into a wonder-medicine for livestock. It is striking that the same pilgrim who holds his audience with a tale of poison is also a vendor of medicine—and that the legend of well water changed to medicine via a Jewish artifact inverts contemporary legends of Jews poisoning wells, thus spreading plague. The double identity of Jews in medieval repute as skilled physicians but poisoners of the water supply, as well as alien bodies in the body politic, is a particularly charged and fateful example of the paradox of the murderous healer.

In *Hamlet*, as the death trap is being set, Laertes mentions that he has in his possession a most lethal poison—obtained not from a doctor but a "mountebank" or vendor:

> I bought an unction of a mountebank,
> So mortal that, but dip a knife in it,
> Where it draws blood, no cataplasm so rare,

Collected from all simples that have virtue
Under the moon, can save the thing from death
That is but scratched withal.

It is hard to imagine such a sinister substance being dispensed by a physician. In *Madame Bovary*, similarly, the heroine finds the fatal arsenic not among the possessions of her own husband (a sort of minor medical practitioner) but in the inventory of the pharmacist Homais, himself a mountebank if only he knew it. She obtains it on the pretext that "she had to kill some rats that were keeping her from falling asleep," which is virtually the same excuse used by the youngest of the three revelers in the Pardoner's Tale. Charles Bovary does not seem like one who would use arsenic. It is one of the lesser sorrows of *Madame Bovary* that he nevertheless seriously violates the rule of not doing harm, the medical precept he was best qualified to keep because he couldn't really do much of anything.

"I will not cut, even for the stone, but I will leave such procedures to the practitioners of that craft," says the Hippocratic Oath. Traditionally there was enough of a distinction between surgery and medicine that the former "lagged well behind medicine in gaining academic recognition," with barbers and itinerants frequently doing the incising. Well into the eighteenth century, barbers and surgeons in England belonged to the same guild. If the progressive Lydgate in *Middlemarch* considers the "severance between medical and surgical knowledge" a vestige of the past, the sort of division of labor envisioned in the Hippocratic Oath, with physicians diagnosing and others cutting, leaves traces in literature of the past. In the *Canterbury Tales* the learned Physician gets rich on the plague (with the implication that his costly ministrations did his patients no good, not that he actually harmed them). Much like today, he has apothecaries at the ready "to sende hym drogges." Although he speaks of "surgerye," there is no suggestion that he actually cuts and every suggestion that one erudite as himself, a man with his head in the stars and an imaginary companion of Hippocrates, Galen, Avicenna, and Averroes, would consider such hands-on work beneath him. In the Miller's Tale Alisoun's unsuccessful suitor, the parish clerk Absolon, lets blood as well as shaving, cutting hair, and drawing up legal documents. We can imagine the Physician delegating his dirty work to a barber-surgeon like this, a mere doer who lacks grounding in theory, much as professors performing public anatomical demonstrations at the time (in Italy) "would sit in a high chair reading out relevant passages from the works of Galen, while [their] assistant pointed to the organs alluded to and a dissector did the knifework." The most comical character in the *Thousand and One Nights* is a babbling barber who does knifework but proclaims himself a man of learning. "Do you want me to shave your head or let blood?" he asks a client. When the client says he wants his head shaved, the other replies characteristically, "You have asked for a barber, and God has sent you a barber who is also an astrologer and a physician, versed in the arts of alchemy, astrology, grammar, lexicography, logic, scholastic disputation, arithmetic, algebra, and history." In a world where one is either a physician or a barber, this prodigy proclaims himself both.

Chaucer's pilgrims are marching toward the shrine of St. Thomas à Becket to offer their gratitude for his help "whan that they were seeke [sick]." Nothing is said of *physicians* helping patients, and certainly the Canterbury Physician did nothing for those who died of the plague all around him, though at least he didn't take to his heels like others of his profession. While the figure of the doctor who harms in the name of healing still haunts our imagination, perhaps the more common satiric image of the doctor is one like Chaucer's Physician who talks a good game but accomplishes nothing. "Don't even think about doing harm," we say to doctors, "for if we satirize you this sharply now, when you injure only our pocketbooks, how might we picture you *then?*"

Although one of Poor Richard's sayings, "He's a Fool that makes his Doctor his Heir"—an observation that also appears in *Volpone*—speaks to our fear of the murderous healer, comments on the uselessness of medicine and the virtues of temperance and moderate diet seem closer to the note and spirit of Poor Richard's wisdom. "He's the best physician that knows the worthlessness of the most medicines." (Signs of therapeutic nihilism.) Just as, in Chaucer, St. Thomas cures the sick and the Physician receives their money, so "God heals, and the Doctor takes the fees." By contrast with the vanity of medical measures, "Time is an herb that cures all diseases." A Bostonian contemporary of Franklin's observed, in the same spirit, that nature is the best doctor (a bit of common sense that evolved into the nineteenth-century doctrine of the *vis medicatrix naturae*, the curative power of nature), and that "frequently there is more danger from the physician than from the distemper." While Poor Richard's proverbs certainly spring from the culture of puritanism, the distrust of medicine they radiate belongs to no single sect, school, or even time. In the opening words of Tolstoy's fable "The Prayer," the mother, distraught over the illness of her child, asks, "Doctor, is it true that we can do nothing? Why are you silent?" The physician can do nothing, though neither does he contribute to the mother's sufferings. He is simply irrelevant. In Tolstoy's dark novella of medical futility, *The Death of Ivan Ilych*, the physicians who purport to treat the dying man are impostors, pretending to know all and in fact knowing nothing, and pretending to do something but accomplishing little. By the 1850s, some thirty years before events in this tale take place, "new, chemically pure drugs could . . . be injected by hypodermic syringe straight into the bloodstream. Morphine was extracted from opium, quinine from cinchona, but people went on dying, more or less as before"—just as Ivan Ilych's disease runs its course despite the injections he receives and despite the scientific airs of those who treat him. But at least these professionals do nothing to hasten his death or aggrieve his bodily suffering. Despite their indifference to the patient as if they belonged to some higher category of being, even these men, somewhere in their being, seem to know the rule Do No Harm.

2. MEDICAL RECKLESSNESS: A FICTIONAL CASE

For years DES was prescribed by obstetricians despite no evidence of effectiveness and a clinical trial pointing to its ineffectiveness, and it continued to be prescribed in Europe for some time even after being banned as a pregnancy drug by the FDA in 1971. Those women who took the drug now confront the possibility that its toxic effects extend even to the third generation. Looking back on the DES story, three researchers recently concluded that the marketing of DES

> had enormous economic advantages for the [pharmaceutical] industry. It also seems plausible that patients took more for granted than they would do nowadays, and that it was more difficult for physicians to stay well informed about the latest developments. All of these factors might have contributed to the fact that, in some countries, DES was still administered years after the FDA withdrew the approval for DES usage. Although it seems unlikely that such mistakes will be made nowadays, it is important to remain focussed on the effectiveness, and the toxicology, of any medication that appears on the market.

Although the twice-used expression "nowadays" seems to say that we are not as blind as those who went before us, we can be sure that many who prescribed and took DES beginning in 1947 would have felt the same way, living as they did in the afterglow of the production of penicillin and in the midst of a surge of new drugs as a result of a long-awaited therapeutic revolution. As the authors themselves recognize, the belief that we stand above our predecessors, exempt from their delusions, walking in light where they walked in darkness, is a dangerous delusion in its own right.

The notion that tradition is a bundle of fallacies and prejudices, a shackle on the human mind, an offense against all that we believe in nowadays, against our very capacity for progress—this notion, now so widely held, and so ardently disseminated even in our institutions of learning, derives from the Enlightenment critique of the authority of the past in the name of reason. In the first instance it was the investigation of nature that provided the model of reason in action, but the prestige of science spilled over into the investigation of society, and the reform of knowledge into reform of the world. As the campaign against delusion, error, and merely traditional knowledge acquired the power of a tradition itself, so too did the idea gain ground that our very institutions must stand the test of reason and, if necessary, be reconstructed under its guidance. Writes Edward Shils in his masterful study, *Tradition*,

> There was a vast thicket to clear but the great men of the Enlightenment and of nineteenth-century liberal scientism looked forward with confidence to a time when much of the area would be cleared. Then reason and scientific knowledge would reign and mankind would not longer be enslaved by tradition.

Jeremy Bentham was one of those "crusaders" (as Shils names them) who applied their energies to clearing away everything obstructing the light of reason, though it needs to be said that crusading is not exactly a rational activity.

At some point the idea that the minds of one's fellow creatures need to be emancipated from thralldom to the past became as well established, in some circles, as deference to the past had once been. Typical of one in whom the Enlightenment ideal of reason has become purely traditional, writes Shils, is Homais, the unforgettable pharmacist-journalist and epigone of Voltaire in *Madame Bovary*. M. Homais talks like this:

> "I believe in a Supreme Being, in a Creator whoever He is, it doesn't matter, who placed us here below to fulfill our duties to family and state, but I don't need to go into a church to kiss silver plates and fatten up a bunch of fakers who eat better than we do! . . . God, for me, is the God of Socrates, Franklin, Voltaire, and Béranger! I am for the *Savoyard Vicar's Profession of Faith* [a

reference to Rousseau's *Emile*] and for the immortal principles of eighty-nine! So I cannot accept a doddering deity who parades around in his garden with a cane in his hand, sends his friend up a whale's belly, dies with a shriek, and is resurrected three days later. These things are completely absurd, and besides, they're completely opposed to the laws of physics."

Thus Homais, a voice of the most resounding unoriginality.

In the spirit of prevention, Homais not only advocates vaccination to the world but tries to keep accidents from happening to his children.

> The knives were never sharpened, floors never waxed. There were iron bars on the windows and heavy guards at the fireplace. The Homais children, despite their independent ways, could not budge without someone to keep an eye on them.

Somewhat like Bentham's imaginary inmates, the Homais children live in a safe, well-guarded place under the watchful eye of benevolence. Homais is less careful of other lives than of his children's, however.

In the town in which Homais as well as Charles and Emma Bovary live, there lives also a stable-hand with a club foot, Hippolyte Tautain. Less for Hippolyte's benefit than for the glory of the town, the journalistic riches to be reaped, and the triumph of demonstrating that only science, not religion, can heal the lame and the halt, Homais proposes to Charles Bovary—not a medical doctor per se, but a semi-accredited health officer—that he perform a new operation to cure the stable boy's deformity. In point of fact, the correction of club foot by sectioning of the tendon marked the beginning of orthopedic surgery. In the context of a novel that satirizes the dissemination of ideas and fashions into the provinces, however, the apothecary's vision of a surgical triumph reads like a parody of contemporary surgical fads, such as "correcting" stutters by cutting the root of the tongue. For an enthusiast of received ideas like Homais, such a craze is irresistible.

"After all, what's the risk?" he asked Emma. "Let's see"—he ticked off on his fingers the advantages of such an attempt—"an almost certain success, relief and improved appearance of the patient, and a rapid rise to fame for the surgeon . . . and besides," here Homais lowered his voice and looked around, "what's to stop me from sending a short article about it to the paper? And, by God, an article gets around; people talk about it; it begins to snowball! And then, who knows?"

A slow man and by no means a brilliant practitioner of medicine, Charles Bovary nevertheless allows himself to be persuaded, and in short order the operation is performed. "The tendon was cut, the operation finished. Hippolyte could not get over his surprise; he leaned over Bovary's hands to cover them with kisses." To protect the foot during his recovery, the patient is outfitted with a heavy wooden box constructed by a carpenter with the aid of a locksmith, and it is in this wretched state that he endures the terrible results of a medical experiment gone wrong. For even while Homais writes in the Rouen *Beacon,*

> "Despite the network of prejudices that still covers part of the map of Europe, light is nevertheless beginning to penetrate our countrysides. And so, last Tuesday, our small city of Yonville was the scene of a surgical experiment that is, at the same time, an act of great philanthropy. . . . The operation . . . was carried out beautifully,"

Hippolyte is suffering from gangrene. A renowned surgeon is sent for, and the victim's leg is amputated—without anesthesia. Probably only because the operation does not take place in a hospital, where mortality as a result of infection runs sky high, he survives.

On the mantelpiece of the Bovary parlor, between two silver-plated candlesticks, stands a clock "with a head of Hippocrates." If only Charles had heeded the famous precept attributed to the same Hippocrates—Do No Harm—he would never have been drawn into performing an operation about which he knew nothing beyond a few scraps of information picked up from a book borrowed for the occasion; but the clock on the mantel inspires no more thought than the wallpaper behind it. The principle of nonmaleficence lives on in Charles Bovary only as a fear of exceeding the narrow limits of his competence and understanding, a modesty that serves him well most of the time.

> The country people adored him because he wasn't conceited. He would pat the children, never entered a café, and otherwise inspired confidence by his good morals. . . . Being quite afraid of killing his patients, Charles hardly prescribed anything but sedatives, occasionally an emetic, a foot bath, or leeches.

There is a certain unconscious wisdom in Charles's confining himself to such modest practices, but not enough wisdom to hold out against the combined importunity of his wife and the ideologue-pharmacist with their respective visions of glory. As Charles gets ready to perform the surgery, "he was even trembling already, afraid of interfering with some important area of the foot that he did not know." His trembling is the last stirring of wisdom before he proceeds to commit, for the lightest of reasons, a terrible and irreversible medical error.

To Homais and Charles Bovary—the pseudo-medical apothecary and the quasi-medical health officer, the zealot of received ideas and the man of no ideas—Flaubert may be saying, "A plague on both your houses"; but at least the health officer is unconsciously in touch with the precepts of caution, precepts that would have saved his patient's leg if only he had remained loyal to them. The pharmacist, sworn enemy of charlatanism, is a charlatan himself. Sworn enemy of tradition, he intones phrases as ritualistic as any catechism, celebrating the operation on Hippolyte as if both the immortal principles of eighty-nine and the advances in scientific medicine following the French Revolution were being played out all over again. And although one of his heroes is Voltaire, never does he stop to wonder what it might mean that Voltaire followed philosophically after a physician—Locke. ("After so many deep thinkers had fashioned the romance of the soul, there came a wise man who modestly recounted its true history: Locke has unfolded to man the nature of human reason as a fine anatomist explains the powers of the body.") Locke's sensitivity to the limits of human knowledge is at one with his approach to medicine, but in any case Locke was still a traditional enough practitioner of medicine to think that medicine was concerned with persons—traditional enough, therefore, not to undertake the curing of a society, as if society were simply a patient writ large. As the language of the Enlightenment became traditional, reformers acting in the name of Lockean principles did just this, playing doctor to the body politic and arrogating to themselves rhetorically something of the authority of science; and of these Homais is an excellent caricature. Enjoying in his capacity as "chemist" a certain ambiguous association with medicine (indeed he once got into trouble for practicing medicine without a license), Homais seeks to exploit the prestige of medicine and of the Enlightenment tradition. But amid all those like him, and they are many, who borrow against the good name of medicine for their various schemes of therapy and reform, medicine retains its own traditions. How ironic that reformers in the line of Homais appeal to the authority of a profession that even today trains its members on the model of apprenticeship.

3. Freud's Harms

It is a disturbing fact that a doctor of medicine widely regarded as the Einstein of the human psyche—the revolutionizer of our understanding of ourselves—rode roughshod over the prohibition of harm. Freud's reputation as a brilliant explorer of the darkness within rests on widespread ignorance of his actual practice. Consider only two cases:

Freud diagnosed one Emma Eckstein as "'bleeding for love' of himself, whereas she was actually suffering from a half-meter of gauze" that his colleague Wilhelm Fliess left "within the remains of her nose" after a quack operation she underwent on the recommendation of both Fliess and Freud. This episode, somehow beyond even the satiric pen of Flaubert, typifies Freud's disregard both of the principle of avoiding harm and the canons of philosophical modesty. It is significant that the two principles stand, and in this case fall, together.

Also characteristically, on the basis of a dubious diagnosis of latent homosexuality, Freud urged a patient, Horace Frink (a psychoanalyst in his own right), to divorce his wife and marry an heiress with whom he was having an affair.

> The divorce and remarriage did occur—soon followed by the deaths of both of the abandoned, devastated spouses, an early suit for divorce by Frink's new wife, and the decline of the guilt-ridden Frink himself into a psychotic depression and repeated attempts at suicide.
>
> It is not recorded whether Freud ever expressed regret for having destroyed these four lives, but we know that it would have been out of character for him to do so.

It seems poetically just that the term "iatrogenic," referring to harms, meaning those caused by physicians themselves, was first used in 1924 in a textbook of psychiatry.

Intentionally or not, Freud did more than anyone to establish the view that the Victorian family was a hothouse of repression and Victorianism itself a system of hypocrisy. And the story of Freud shattering the Victorian conspiracy of silence bears indirectly on the subject of medical harm, for not only did the Hippocratic prohibition come to the fore during the Victorian period, but those who now criticize it as inconsistent with the practice of medicine regard it too as so much sentimental hypocrisy—in effect, a Victorian piety. Like a Freud attacking the refusal to face up to facts, they portray the prohibition of medical harm as *itself* a refusal to face facts. As the facts of Freud's own conduct suggest, however, we discard such a prohibition at our own peril. It is true that information about the harm Freud did will make little impression either on true believers or those who admire Freud as a culture-genius while ignoring his actual methods. Both groups seem to feel that if a few suffered in order that the light of Freudian knowledge might illumine the world, that is a small price. I wonder about the quality of "scientific" knowledge founded upon blind indifference to harm. Is a shattering conceptual breakthrough that also shatters such an elementary principle as the medical maxim Do No Harm worth having?

Psychoanalysis grew up on the kinds of patients once treated by doctors with elaborately innocuous measures—patients like the anonymous narrator of Charlotte Perkins Gilman's now-famous story "The Yellow Wallpaper," who reports, "I take phosphates or phosphates—whichever it is—and tonics, and air and exercise, and journeys." The strange hysterical disorders exhibited by the first generation of psychoanalytic patients were very much in the tradition of other maladies of obscure origin that appeared in certain circles from the eighteenth century on: "the vapors," hypochondria, neurasthenia. They were also in the tradition of that pioneer of hysteria and addict of the imagination, Emma Bovary, for whom no prescription—whether valerian drops, camphor baths, tea, or a change of scene—does any good, her ailment being beyond the reach of medicine. Although it has been said that in the nineteenth century "medical men maintained the charade [of patients with nonmedical illnesses] by cooking up innovations in minor surgery, spa regimes, coloured waters, and health-retreat regimes, relying on the desperate gullibility of patients prepared to believe that each and every organ could spawn dozens of defects," a less satiric interpretation might be that such doctors treated questionably medical ailments within the constraints of the principle of avoiding harm, neither threatening the patient's sense of self by attacking her fictions nor recommending dangerous therapies like arsenic (as one French doctor did in a treatise on hysteria) but instead prescribing what amounted to placebos. In the end these masters of doing little may have been wiser than one who professed to solve hysteria by inventing a fictive system of his own.

4. INFAMY: THE TUSKEGEE SYPHILIS EXPERIMENT

Whether performed by physicians, surgeons, apothecaries, or laymen, with lancet or leech, the traditional curative practice of bleeding appealed to the intuitive belief that getting rid of some bad blood serves to cleanse the system. In the United States the notion of bad blood lasted well into the twentieth century, figuring in the most notorious medical episode in American history, the Tuskegee Syphilis Experiment. That experiment did not begin with the intent of doing harm, but evolved it over time.

Launched in 1932 under the auspices of the United States Public Health Service, the Tuskegee Syphilis Experiment was originally intended as a study of the effects of untreated tertiary syphilis. In the belief that they were suffering from the mysterious ailment known as bad blood, some four hundred infected black field workers from Macon County, Alabama, were followed medically from year to year with a view to claiming their bodies at death for the autopsy table. The object of the experiment was achieved as the subjects died. Harkening back to the practice of treating paupers in order to dissect their corpses, the experiment had the additional stain of deceit: the men were induced to believe that the doctors were interested in their health, when in fact they were interested in their organs. (In the words of the last survivor of the experiment, who died in 2004, "They said it was a study that would do you good.") Once well in the study's net, the men were treated only with iron tonic and aspirin, the latter of which, being new to them, seemed a wonder drug. At the time the study began, the standard treatment of syphilis involved a sort of failed wonder drug, Salvarsan, which was acclaimed as a magic bullet when it was discovered in 1910—thus giving rise to the myth of the same name that persists to this day—but proved difficult to administer, toxic, and dangerous. Retrospective apologists for the Tuskegee experiment argued that Salvarsan treatment brought "more potential harm for the patient than potential benefit." Thus the experiment's crimes may originally have been rationalized by the principle of avoiding harm and by the tradition of therapeutic nihilism now in disrepute but at one time not only estimable but dominant.

Designed as it was not to treat the four hundred subjects but to discover what syphilis did to their bodies, the Tuskegee study brings to mind the nineteenth-century belief that because of the little medicine can actually do, "the real function of medicine was to accumulate scientific information about the human body rather than to heal." (There is something of this in the physicians attending Ivan Ilych in Tolstoy's novella, men who pose as priests of science and lords of the body even as they do little or nothing for the patient.) As ruthless and mendacious as the study was, those who ran it could congratulate themselves on their own rigorous medical realism, born of a distrust of suspect cures and wishful thinking. William Osler himself, the most renowned professor of medicine of his time, and a believer in not doing harm, had written in 1892 that owing to the inefficacy of medicine, many diseases were not to be treated. Moreover, around the time the Tuskegee experiment was conceived, "some studies . . . suggested that syphilis did not always need to be treated—that it could often remain quiescent, especially in blacks." Some years into the experiment, however, the game changed when penicillin, a highly effective treatment of syphilis, came into use. In order to keep the experiment going just when it would seem to have lost its point, a decision was taken to withhold the drug from the men, even to see they didn't get their hands on it, after which the study coasted on its own momentum for another quarter-century. A study of untreated syphilis in Oslo, Norway, tracked the medical records of some hundreds of men from 1891 to the year Salvarsan came onto the scene, 1910: a terminus implying that a study of untreated syphilis comes to an end when a treatment for the disease (and a much-acclaimed treatment at that) becomes available to the world. The Tuskegee study continued for two decades and more after the arrival of a drug that fulfilled the hopes originally raised by Salvarsan, becoming "the longest observational study in medical history." The long shadow of this abomination may have reached the PCPT, which began but twenty years after the Tuskegee study was dismantled, and which succeeded in recruiting only small numbers of black men—4 percent of the study's participants—despite their high stake in prostate cancer prevention. The PCPT's failure to enroll sufficient numbers of black volunteers has been laid in part to the "substantial distrust of government-supported research" that lingers in black America. In 2000 a conference on the underparticipation of black citizens in clinical trials was held in Tuskegee, Alabama.

Clearly the Tuskegee Syphilis Experiment was a case of bloodless mass murder. Thought staggers at the sheer calculated indifference shown both by the medical authorities who originally maintained that "little harm was done by leaving the men untreated" (because no good treatment was available) and by the many in the medical world at large who knew of the experiment and had no objection. But it is a matter of speculation whether the same doctors would have taken part in an experiment designed from the start to keep men with syphilis from a drug that could definitely cure it, and cure it without the rigors, dangers, and expense of Salvarsan. The Tuskegee Syphilis Experiment was not conceived as an exercise in harm. While the view that treating with Salvarsan was worse than doing nothing is open to question, and while the experiment's methods from top to bottom were indefensible, its original rationale of studying disease rather than leaping to cure it would at one time have seemed scientific. And by the time penicillin became available, "more than a decade had passed since the men had been given any form of treatment for syphilis. . . . Not treating them had become routine." At this point those in charge could tell themselves that too much time and effort had been invested to call the study off—there was no going back. Science could not be denied such valuable results.

Would the good doctors have launched a study in which, from the first day, a highly effective cure was systematically withheld from men with a ravaging disease? Not having acquired its own inertia as the Tuskegee study did by the time penicillin came on the scene, such a study might have seemed just too shocking and pointless, too contrary to good practice, to pursue. Doctors not already steeped in the experiment before they began withholding penicillin would be unable, or less able, to claim they would be cheating science of important findings unless they pressed grimly on. (Obstetricians who would never have begun prescribing DES on the strength of evidence that it was useless were able to ignore that evidence once they were in the habit of prescribing the drug.) They would be less able to finesse the question of harm. People can bring themselves, or be led, to do things over time that they would never have done originally. Intentions evolve. No one originally intended the Tuskegee study to run for decades at all. It snowballed, and as it did so, medical authorities who at first thought "little harm" was being done, proceeded to do great harm. Some of the Tuskegee men went blind, some went insane, and many died of a curable disease.

The author of a study of the Tuskegee experiments rightly dismisses medical rationalizations of its crimes, reminding physicians of their duty "to prevent harm and to heal the sick whenever possible," and citing the Hippocratic Oath. Physicians today who swore to a version of the Hippocratic Oath with the prohibition of harm edited out, or who pooh-pooh the prohibition of harm as so much sentimentality, might reflect on the Tuskegee precedent.

5. THE POWER OF SUGGESTION

The Tuskegee story is, among other things, a study in abused trust. Like some drug that can be used either as medicine or poison, trust itself can be cynically exploited or used to good healing effect. Instances of imposture and deception may stand out in our minds, but the lore of medicine is also rich with stories of patients cured by trust in the doctor—that is, by the very hope of a remedy. Such is the power of suggestion, especially over ailments of psychological origin, that the mere concern shown by a doctor attentive and understanding enough to hear a patient out can have great therapeutic benefits. "The healing power of the consultation lies in the catharsis that the patient derives from telling his story to someone he trusts as a 'healer.'" Contrary to the general belief, Freud did not invent the talking cure. Suggestion has long played "an enormous role in the practice of medicine," and the reluctance of physicians today to assume the person and authority of advisers, or to offer placebos, helps explain the flight of patients to psychologists who are ready and willing to do what physicians are not.

But if suggestion exists, if in fact doctors possess such influence that their placebos can cure and their very demeanor can have a therapeutic effect—and to me the evidence points to the conclusion that they do—those doctors who once lied to their patients, fearing that telling the truth would harm them, may have been on to something. After all, why should suggestion work only for good? If sheer influence can help patients, why can it not also injure them? Writes Worthington Hooker (himself a critic of the deception of the ill for their own good), "The cordial influence of hope . . . is often one of the means by which a recovery is effected, and *the absence of this one means may prove fatal*" (emphasis in the original). Toward the end of his survey of the medical advances of the nineteenth century, William Osler launches into a panegyric on the power of suggestion. "Faith in the gods or in the saints cures one, faith in little pills another, hypnotic suggestion a third, faith in a plain common doctor a fourth." He proceeds to caution against the irresponsible use of the third method of suggestion, hypnotism, warning that it is "not without serious dangers, and should never be performed except in the presence of a third person, while its indiscriminate employment by ignorant persons should be prevented by law." But if the hypnotist can harm by the power of suggestion, why cannot plain common doctors do the same? Who is to say that their wariness of causing harm by telling truth was unfounded? Not that we ought to revive, somehow, the days when physicians lied to patients for their own good. The nurse who acted as an intermediary between the masters and the subjects of the Tuskegee study defended the denial of treatment to the latter with the argument that they received the full blessings of the placebo effect. "They didn't get treatment for syphilis, but they got so much else. . . . They enjoyed having somebody come all the way from Washington . . . and spend two weeks riding up and down the streets looking at them, listening to their hearts and [having] somebody to take their blood pressure. . . . That was as much help to them as a dose of medicine."

But the placebo effect need not be employed cynically, and in fairness to the past, and despite our belief that paternalism was a complete sham, we should consider that some lies to the sick may really have been well intended. In any case, a physician today who made it a principle not to lie to patients—one of the lessons of the Tuskegee infamy—but who had neither the time to hear their story nor the inclination to dabble in suggestion, would be in a poor position to criticize those predecessors who at least knew the power of suggestion and sought to use it well according to their lights.

[10]

"Do *No* Harm?"

◦◯◦◯◦◯◦

THE STORY GOES that a man fell dead of a heart attack on a busy street. As he was being tended by a policeman, a matron of the Jewish faith strode up, saying, "Give him some chicken soup!" "But madam," says the policeman. "He's dead. What good would chicken soup do?" Says she, "It wouldn't hurt."

1. HARM AND DREAD

A physician critical of the prohibition of harm might say, "Do *no* harm? Surely you don't mean that. Are we to rely on the medicinal powers of chicken soup? If I am to do no harm—no harm whatsoever—I couldn't prescribe tamoxifen for breast cancer patients even though it will prevent thirty times more deaths than it will cause. In the face of those numbers, a doctor who refused to prescribe tamoxifen could take pride in being a purist maybe, but he would certainly not be doing his patients very much good. Human life, and the practice of medicine in particular, don't allow for such absolutes as the rule Do No Harm." Perhaps; but that is not to say that the rule has no force. In 1992, before the initiation of the PCPT, a panel on the chemoprevention of prostate cancer laid down the principle that

> where *high toxicity* is acceptable to treat an **established malignancy**, only a *slightly elevated degree of toxicity* is tolerable and acceptable if the target population is at **elevated risk**. . . . Populations with an **intermediate risk** of developing cancer would require *even less toxicity*. To reduce overall risk in large populations, chemopreventive agents must be entirely free of any side effects. [emphasis in the original]

This, I believe, is the sort of reasoning that has ruled out the general use of finasteride, whatever critics might say about the impossibility of any drug other than a placebo being free of toxic effects. To administer a risky drug to a large population over a long time is asking for trouble.

It is worth noting that the revolution in prevention that virtually divides the history of medical practice in two—Lister's introduction of antiseptic surgery—was inspired by the thought of killing the agents of infection without harming the patient. As Lister later said, "Just as we may destroy lice on the head of a child . . . by poisonous applications which *will not injure the scalp*, so, I believe, we can use poisons on wounds to destroy bacteria *without injuring the soft tissues of the patient*." Temporally closer to the issue at hand is another advance in prevention, the most significant in recent decades—the reduction in smoking in token of the link between smoking and lung cancer established in 1951 by Richard Doll and A. Bradford Hill. The cessation of smoking does not cause harm—and the benefits of not smoking dwarf those of finasteride, even in optimistic projections. An analysis of the PCPT, seeking to place the putative benefits of finasteride "in perspective," notes that

> A program of physical activity begun at age 35 increases life expectancy by 6.2 months. Complete smoking cessation at age 35 increases life expectancy by 9 months. Childhood vaccines

against measles, rubella, and pertussis, however, increase life expectancy by about .1 month each.

Under no model, including those that assume no increase in high-grade cancer, does the projected use of finasteride yield survival benefits on the order of nine months. Again, it has been calculated that so modest a measure as "a nation-wide moderation of salt intake" could prevent a quarter of the strokes and a fifth of the heart attacks in the United Kingdom. Although food might taste a bit bland at first, I think no one would classify that as a positive harm.

Admittedly the conditions of life are such that at times we have no choice but to transgress rules and principles as primary as the medical prohibition of harm. "Thou shalt not kill," says the commandment, and yet in war we do kill (a contradiction that for some becomes unbearable). But if in exceptional circumstances we violate the prohibition against killing, this doesn't mean that "Thou shalt not kill" is nothing but a bit of unrealistic rigidity or a pious fraud we would do well to wipe from the books, to be replaced with something less demanding, such as "On the whole, do not kill." The author of a study of murder in early-modern England reminds us that "95% of unmarried mothers did not kill" their infants, implying that an irrational society victimized the 5 percent who did. "Contemporaries may have feared them and prosecuted them in relatively high numbers, but these women were never a real source of absolute danger in terms of the number of killers and the numbers killed." It seems early-modern England should have looked at the numbers, taken comfort from the finding that on the whole women didn't kill their infants, and recognized that, realistically speaking, a minority of murderers in no way threatens the general welfare.

Some medical modernists might prefer the accommodating maxim, "On the whole, do no harm." As noted, a popular revision of the Hippocratic Oath simply edits out the prohibition of harm, quietly depriving the traditional oath of its most salient feature even while embellishing the original with vaguely uplifting exhortations, such as "I will remember that I remain a member of society, with special obligations to all my fellow human beings." Explicit critics of the Hippocratic maxim point out, too, that doctors do not have the moral luxury of doing nothing whenever the risk of harm presents itself. As one contends,

> although the rule "above all, do no harm" cannot be logically sustained, "never forget the possibility that you may do more harm than good" makes good sense. Though lacking in neatness and brevity, it could no doubt be rendered in Latin for those who love the quasi mystical authority of an ancient language.

But if "almost every effective treatment can sometimes do more harm than good," as the same critic alleges, a conscientious doctor will never forget the risks and go ahead and do whatever he or she intends to do anyway. With a snap of the fingers, Do No Harm is replaced by a piece of advice which in turn instantly reduces to nothing. But it bears noting that this critic indeed has treatment in mind—not prevention. He regrets that "adverse drug effects . . . get far more publicity than do deaths or disabilities due to lack of sufficiently aggressive treatment," and remarks that "the gut feeling of many people is to be especially unhappy about any death or disability that is caused by treatment," as if this were one more lamentable display of the illogic of the human animal. Yet if it is difficult enough to weigh the harms against the benefits of treatment (as the critic himself seems to acknowledge), how much more difficult is it to compare the possible harms and the speculative benefits of prevention? It is one thing to treat a patient with a disease, but quite another to prescribe a chemo-preventive drug to patients who may never develop the disease in question, or whose disease might not prove serious even if it did come about—or, alternatively, might be aggravated by the drug itself, as in the case of finasteride.

It is true that "certain kinds of iatrogenic injuries do not result from errors at all—for example, predictable side effects of drugs which are utilized in full awareness of their side effects because, all things considered, the decision to accept the side effects constitutes the best available response to the presenting symptoms." The treatment of prostate cancer stands as a particularly embarrassing example of injury inflicted by the standard of care itself, as it causes not only harms to the patient but harms possibly worse than the original disease. In part, the hope of breaking the intolerable dilemma of prostate cancer medicine and relieving the patient of the burdens, costs, and griefs of overtreatment is what makes the prospect of chemoprevention so appealing. As the National Cancer Institute stated in its announcement of the PCPT findings in June 2003, "the disease [of prostate cancer]—as well as its treatment, which sometimes leads to impotence, urinary incontinence, and other problems—causes a significant health burden for men." But note that the preventive use of finasteride doesn't fit the model of a standard treatment that harms, just as the drug itself is not one which, for all its side effects, "constitutes the best available response to the presenting symptoms." For one thing, there are no presenting symptoms. *The fact is that the chemoprevention of prostate cancer introduces such imponderables as to defeat the very model of the balance scale weighing risks and benefits.* Finasteride turned into a quandary, or a muddle, precisely when this medication used for treatment became a candidate for use in prevention, as if the same sort of balancing of risks and benefits that is done in the first instance could be done in the second. Consider again the observation of a European team reviewing the findings of the PCPT: "the potential mortality resulting from one high-grade tumour might equal that from several of a lower grade, and a greater incidence of more aggressive disease could outweigh any benefit from an overall reduction in cancer risk." A scale that gives such indeterminate readings no more resembles a conventional instrument than Salvador Dali's melting watch resembles a conventional timepiece.

Critics of the Hippocratic rule regularly appeal to the model of the balance scale:

> To observe this advice [that is, Do No Harm] literally is to deny important therapy to everyone, since only inert nostrums can be guaranteed not to do harm. It is more reasonable to ask doctors to balance the potential gains against the possible harm; would that we could only quantify these probabilities more precisely!

But imagine a drug that would spare exactly 4.6 or 8.2 or 10.9 or 16.3 cases of prostate cancer for every case it made more dreadful. Do these exact numbers and rising ratios in any way change the picture? The balance scale of the finasteride literature doesn't even perform the function of a scale. According to the follow-up paper on the PCPT that supplies the figures just used, in calculating the harms and benefits of finasteride "every individual will assign his own weights" based on age, risk, sexual factors "and other indefinable values." What kind of scale is it that allows the user to choose how much something weighs? The balance scale is an overrated model. Some things that need balancing cannot be reduced to quantities—ideologies, for example. The Hippocratic rule rose to prominence when physicians sought some intelligent mean, some balance between the ideologies of letting Nature heal itself on the one hand and attacking disease with anything and everything on the other, leading to harms of omission and commission, respectively. Many hope that finasteride will solve the corresponding urological dilemma of doing nothing ("watchful waiting") on the one hand or treating aggressively on the other. As things stand, however, the Hippocratic rule argues against throwing finasteride at prostate cancer.

Critics may debunk the Hippocratic rule, or mock it as a superstition or an archaism, or simply write it off, yet most nevertheless carefully avoid the issue of deaths caused by medical error, as if they dreaded it—as well they might. Not everyone in medicine dreads it. To crack jokes about fools who get especially unhappy about deaths attributable to medical treatment is to say that physicians needn't dread causing death. Some in the medical world and even some ethicists, evidently moved by the spirit of pop psychology, would like to soften the interdiction of harm to allow for the blame-free reporting of medical errors, up to and including those leading to death. In a commentary entitled "When Primum Non Nocere [First Do No Harm] Fails," the editors of *The Lancet* argued a few years ago that the aura of dread and disgrace surrounding medical errors merely serves to keep the problem hidden. By their own figures, the problem in this case consists of at least 44,000, and possibly as many as 100,000, deaths in the United States per year as a consequence of medical error, a number exceeding the annual toll of motor vehicle deaths or deaths from breast or prostate cancer. It seems to me that such numbers very well ought to evoke dread; that physicians and others who cause or contribute to the death of patients ought to feel that the last line was crossed and not that some anonymous instance of system failure took place; that pop psychology, with its puerile ridicule of the blame game, has no place in a matter of such moment and magnitude as this; and that when physicians or nurses kill patients, it is not the case that the principle of Do No Harm failed them, as the title of the editorial misleadingly suggests, but that they failed it.

Some in medicine go so far as to argue, or at any rate imply, that the rule Do No Harm is itself a cause of great harm, responsible in fact for the epidemic of deaths by medical error. Using a sort of reverse logic, such critics allege that it is because physicians feel too much responsibility, not too little, that so many patients are lost to medical error, and that if only the oppressive burden of guilt and blame were lifted from the physicians' shoulders the entire system would be much healthier and safer. For example:

> The "do no harm" concept amounts to a form of denial, which paradoxically can increase harm. Rather than "embracing error" so as to learn from it and reduce it, we operate under a form of self-delusion that errors don't or shouldn't exist. So errors tend to be ignored, rationalized, minimized, denied, and presumably repeated—as the Institute of Medicine report [finding at least 44,000 deaths by medical error annually in U.S. hospitals] asserts.

If this argument does not sound strange, it is because catchwords like "denial" have already been carried across the length and breadth of the land on the wings of the pop psychology movement. Doesn't pop psychology crusade against obligation and prohibition, and especially against "perfectionism"—what the critic calls the delusion that errors shouldn't exist? As I show in *Fool's Paradise*, my study of the movement, pop psychology everywhere intimates that only with the abolition of morality and everything associated with it will human beings at last behave morally. (The medical version of this abolitionist argument would be: only with the lifting of the prohibition against doing harm will medicine cease doing so much harm.) As I also show, this sweeping utopian proposition is a traceable legacy of the 1960s revolt. Pop psychology's allegation that morality is immoral, its preaching against preaching, its goal of replacing our wiring for misery with a program for happiness, its love of reverse logic, all grow out of a movement whose ambition it was to emancipate humanity by overturning constraints and prohibitions. The argument that the prohibition of harm causes incalculable harm also turns things on their head. Is it wise to annul the first principle of medical practice on the strength of the visionary expectation that doing away with responsibility will make for better doctors? (If indeed "most preventable [medical] harms are the result of system failures" and not the deeds or misdeeds of a person, then maybe responsibility is too dilute already.) Note that the physician quoted, while arguing that the prohibition of harm "paradoxically can increase harm," offers no evidence of any kind that abolishing the hated precept can, does, or will result in better medicine. That happy outcome is simply preordained, like the happy ending of the generic life-stories in the pop psychology section. We do not know the future, but we know something about the past, and it was only when medicine recognized its own responsibility for the transmission of fatal infections that antisepsis won out. The rule Do No Harm, now criticized by some as a fallacy and a danger, underwrote the most dramatic gains in safety in the history of medical practice.

The figure of a minimum of 44,000 deaths per year in the United States as a result of medical error derives from a report published in 2000 by the Institute of Medicine, entitled *To Err Is Human: Building a Safer Health System.* Here too it is asserted as an inarguable principle that blame is a backward practice that obstructs the improvement of medical care. "Building safety into processes of care is a more effective way to reduce errors than blaming individuals. . . . The focus must shift from blaming individuals for past errors to a focus on preventing future errors by designing safety into the system." We are to believe that, as a rule, assigning responsibility even for lethal errors is a pointless exercise. The authors agree with other critics of the existing system that medicine should forgo perfectionism and learn from its mistakes (for, after all, to err is human), but unlike them they are not willing to see their argument through to the point of rejecting the Hippocratic rule itself as unrealistic and harmful. In fact they cite it prominently in their preamble. "'First do no harm' is an often quoted term [sic] from Hippocrates." But if they were really committed to their argument that blame distracts from the task at hand (namely, fixing the system), they too would reject the prohibition of harm, since without blame attaching to the violation of a prohibition there is simply no such thing as a prohibition.

Because they seem to recognize that abolishing the prohibition of harm would not be a good idea, they espouse it (indeed, the phrase floats on the book jacket) even though it contradicts some of their own axioms. In all, their message boils down to: Do no harm, but if you end up killing someone, remember "to accept error as normal" and embrace it "as an opportunity to learn." Compared to this bit of doubletalk, the principle Do No Harm possesses an admirable clarity.

Confronted with the disturbing fact of medically caused deaths numbering somewhere in the tens of thousands annually—but exactly where in the tens of thousands it seems impossible to say—an observer might conclude not that the medical profession was the victim of its own perfectionism but that it was effectively cloaking its actions from public view. It seems that in some degree it still follows the original (1847) American Medical Association Code of Ethics prohibiting one physician from publicly criticizing another, to the end of protecting the reputation of the fledgling profession and setting medicine as a profession above the general fray. (There was some closing of ranks in the British profession too. "It was frequently said that many deaths due to chloroform were hushed up or not reported" in Britain.) What is doubly shocking about the finding of some tens of thousands of medically caused deaths per year is that this was not the first time figures of that order of magnitude had come to light. A quarter-century before the study in question, in 1974, a Senate investigation found that some twelve thousand Americans died each year as a consequence of unnecessary surgeries alone. That it took a high-powered investigation to unearth a crime of this magnitude suggests that many in the medical profession took part in concealing it. That the deaths resulted from unnecessary surgery surely explodes the claim that it is the very dread of doing harm that somehow underlies the epidemic of medical harm. Doctors who perform gratuitous surgery dread causing harm too little, not too much.

Three years before the Senate investigation, the FDA finally got around to banning diethylstilbestrol as a pregnancy drug; recall that it had been in obstetric use since the 1940s despite no evidence of its efficacy, considerable evidence of its inefficacy, and concerns about its dangers virtually from the beginning. Two historians conclude their searching review of the DES record with the words, "Perhaps the ancient Hippocratic injunction, 'do no harm,' need not yield so easily to the demands of 'modern' medicine." If doctors today recommended finasteride for all men over fifty, and if it turned out that it did in fact induce high-grade cancer just as the PCPT data suggest (but now in tens of thousands of men), does anyone doubt that historians, looking back at this misery, would ask why medicine defied such a well-established and elementary precept as Do No Harm?

The sociologist of culture Philip Rieff has written, "Every culture is so constituted that there are actions one cannot perform; more precisely, would dread to perform," and these interdictions are engraved in our character. As we know, there are those both inside and outside medicine who seem to think that doctors internalize the standards of their profession, and the corresponding dread of transgressing them, too deeply for their own and the profession's good. According to one critic of Hippocratic ethics, the imperative Do No Harm "is increasingly subverted, ignored, altered, reinterpreted" in practice, quite as if those concerned simply didn't feel the prohibition against doing harm *as* a prohibition. I would find it impossible to trust someone who played like a lawyer with the cardinal precept of medicine, now disregarding it, now twisting it, now mocking it as a sentimental archaism, now paying it lip service, now blaming it for the scandal of medically caused deaths, now taking care not to mention those deaths at all. John Stuart Mill once sketched the sort of character implied in Bentham's vision of the human being: "Man is conceived by Bentham as a being susceptible of pleasures and pains, and governed in all his conduct partly by the different modifications of self-interest . . . partly by sympathies, or occasionally antipathies, towards other beings. And here Bentham's conception of human nature stops." Among those regions of human character Bentham seems to have no idea of is the core of our being where interdictions are engraved: conscience. "Nothing is more curious than the absence, in any of [Bentham's] writings, of the existence of conscience. . . . There is a studied abstinence from any of the phrases, which, in the mouths of others, import the acknowledged existence of such a fact." Along with "conscience," Bentham writes around words like "principle" and "duty," shunning them like so many old wives' tales or mystifications. The finasteride papers write around the interdiction Do No Harm in the same way. It is never cited or spoken, which does not, however, mean that the authors are prepared to disown it.

Indeed, the force of this principle is still everywhere felt in the literature, and if the authors do not recommend the general use of the drug despite its impressive statistical showing, and despite the urgency of the need to prevent prostate cancer, it is only because they remain mindful of the duty not to harm. Similarly, despite talk of a paradigm-shift toward population medicine, the profession is hardly prepared to abandon medicine centered on a patient and the traditions associated with it. And despite bold medical criticism of the injunction to do no harm, few are prepared to justify medically caused deaths with equal boldness, as a mere side effect of the practice of modern medicine—the price of progress. The case of Dostoevsky's famously torn intellectual, Ivan Karamazov, is instructive. Although Ivan espouses in theory the radical maxim "everything is permitted," he is not prepared to act on it and is undone when another does act on it in his (Ivan's) name. Ideas such as the obsolescence of traditional medicine float through medicine, but this does not mean that physicians by and large are ready to burn the book of the past, any more than Ivan is.

In exceptional circumstances physicians may claim justification for doing harm, which does not amount to arguing that the rule Do No Harm shouldn't exist. "The moral obligation to do no harm can be justifiably overridden but it can never be erased." According to the authors of this admonition, in order for a medical harm to be justified, each of two conditions must be met. "The first is consent by the informed, competent patient (or valid surrogate). The second is that the anticipated harm is necessary to achieve patient benefit and is proportionately less harmful than the condition for which the patient sought care." In the present state of knowledge, the general use of finasteride for purposes of prevention would seem to infringe both of these conditions, especially the second. As I have argued, the actual workings and effects of the drug are matters of such uncertainty and dispute that consent to take it could be considered "informed" only in some minimal sense of the word. More decisively, anticipated harm in the form of a significantly higher rate of more aggressive cancer is not proportionately less injurious than the disease being prevented, for most cancers of the prostate are not aggressive, and, as we know, it may be that finasteride suppresses less dangerous only to promote more dangerous malignancies. But observe that the framers of the two principles just cited do not have in mind cases where a drug is given speculatively to an entire population, like raffle tickets dropped over a city, with the expectation of producing benefits here and injuries there. Regardless of the particulars of the drug's effects, they might well consider such conditions of usage unethical. If, as other commentators have written, "it seems more just, when one knows that someone will be victimized by a program [as with adverse reactions to a vaccine], to construe the costs of caring for the victim as part of the cost of that program, than to let the financial burdens of such care fall randomly where they may," how just is it to let different grades of cancer fall where they may? That human hands have nothing to do with the final distribution of misfortune in this case means nothing. Justice is blind, but that doesn't make a blind undertaking just.

2. WE AND THE VICTORIANS

Struck by the general anxiety over health today despite the advances of medicine and the improvement of health itself, a thoughtful student of medical history concludes that

> much of the loving support that men and women once gave each other in the—now much despised—"Victorian family" helped reduce their anxiety about symptoms. It is the withdrawal of that support, as the nuclear family comes crashing down, that leaves us confused and frightened at the signals of our own bodies.

The same historian makes the case that doctors no longer enjoy the trust they did in days when their manner of listening, their mere concern, even their placebos, had a healing effect. People now go to psychologists not only to relieve the confusion and fright that their forebears may have been spared but to be listened to and to receive concern and placebo-like messages. Even as we pride ourselves on our rebellion against the past, and in particular the Victorian past—the birthing hour of modern medicine— in some sense we seek what the Victorians had.

Before we came to see the present as a bridge to the future, our predecessors the Victorians did. "The one distinguishing fact about the [Victorian period]," writes Walter Houghton in his survey of the era,

> was "that we are living in *an age of transition*." This is the basic and almost universal conception of the period. And it is peculiarly Victorian. For although all ages are ages of transition, never before had men thought of their own time as an era of

change *from* the past *to* the future. Indeed, in England that idea and the Victorian period began together.

So too was medicine in transition, from the clash of competing sects to the supremacy of a single profession; from such conflicting practices as nonintervention on the one hand and the free use of arsenic and mercury on the other, to the more cautious, scientific, and regulated practice known to us today. The tale of doctors unwittingly killing their own patients through infection, and belatedly coming to recognize the blood on their hands, and transforming their practice and their understanding in consequence—as happened in England over the latter third of the Victorian century—is a Victorian story line in itself; and it is as a result of this massive drama that medicine as we know it took shape. Before Pasteur gave Lister the clue to sepsis, Oliver Wendell Holmes in his paper on "The Contagiousness of Puerperal Fever" indicted the "irreparable errors and wrongs" of physicians. "God forbid," he wrote, "that any member of the profession to which [a woman giving birth] trusts her life, doubly precious at that eventful period, should hazard it negligently, unadvisedly, or selfishly!" Evidently Holmes would not have considered the Hippocratic rule moonshine. Like the control of puerperal fever, anesthesia too dates from the Victorian era, if not precisely from Victorian England. Among the first to activate the medical principle of avoiding harm were physicians like John Snow who used chloroform in thousands of cases without a fatality. Finally, only such demonstrations of safety could quell the objections of those who for whatever reasons opposed the introduction of anesthesia, especially in childbirth. (Snow himself attended Queen Victoria in childbed on April 8, 1853.) John Stuart Mill affirms the beneficial effects of the clash of ideas. It may have been the entrenched opposition to chloroform that inspired or compelled its advocates to perfect its use and thereby raise the standard of medical practice. In the case of finasteride, we still place the burden of evidence on the advocates of innovation, though the issues are argued out in the pages of the medical literature without the sort of invective freely employed in the chloroform debate by the supposedly more repressed Victorians.

Almost concurrently with the anesthesia controversy, the ancient maxim, Do No Harm, came back to life. Its first appearances in print, both in the United States and Britain, date from the mid-nineteenth century—Florence Nightingale, for example, writing that "It may seem a strange principle to enunciate as a first requirement in a hospital that it should do the sick no harm." While the explicit citations in print seem to be few, this may mean only that few were willing to state in so many words a principle capable of indicting so much of the existing practice of medicine. The discovery by Ignac Semmelweis in 1847 that doctors themselves were responsible for puerperal fever entailed "an excruciating accusation." (Little wonder that around mid-century many doctors in Britain were reluctant to report patients who died of puerperal fever. They must have known they themselves would be looked upon as the cause of death.) Two years later the American Worthington Hooker wrote in his masterly *Physician and Patient*:

Heretofore the great object of the physician has been to do *positive good* to the patient—to overcome disease by a well-directed onset of *heroic* remedies—and it has been a secondary object altogether to guard against doing him harm. But medical practice is becoming reversed in this respect. It may at the present time be said of quite a large proportion of the profession, that it is the principal object of the physician to avoid doing harm to the patient. . . . "The golden axiom . . . that it is only the *second* law of therapeutics *to do good*, its *first* law being this—*not to do harm*—is gradually finding its way into the medical mind, preventing an incalculable amount of positive ill." So remarks Dr. Bartlett in a work, which I deem to be one of the best and most effectual efforts, which have been made in promoting the revolution, which is now taking place in the practice of medicine.

At the time the rule against doing harm, far from being innocuous itself, conveyed the force of a condemnation. One reason for the resistance to Lister's theory and practice of antisepsis must have been that some physicians simply could not bring themselves to accept guilt for infecting their patients, especially when a supremely exonerating explanation of the high rates of hospital infection was available: that oxygen itself was the culprit. "Imagine what it must have felt like for . . . a surgeon to accept a theory that confronts him with the intolerable fact that for the fifteen previous years of his career he has been killing patients by allowing into their wounds microbes which he should have been destroying." Dismissed by some today as a sleepy piety, the rule Do No Harm was in 1870 a summons to moral clarity and courage. At some point, perhaps by the turn of the twentieth century, some version was in common medical use.

What does it mean that medicine owes, if not the rule itself, at least its salience, to the Victorians? Associated as it is with sexual repression, a colonial empire, and a homely, long-lived queen, the Victorian period has for decades been regarded not only with distaste but with a unique, almost personal antipathy. Modern art swept away the Victorian figure, Freudianism the code of Victorian reticence, modern architecture the existing rhetoric of public buildings, modern prose the prolixity, chattiness, and sentiment of its Victorian predecessor. In view of the backlash against the entire Victorian era, to the point where the word "Victorian" has become an epithet, it would be surprising if a medical legacy of that era did not come in for both criticism and satire. Among the most disreputable qualities of the Victorians is their sentimentality, and critics of the prohibition of harm contend in effect that it too is nothing but a piece of sentimentality— "a rule of thumb that [feels] good" but doesn't stand up to the test

of reality, just as one might say of a novel that gratifies our sentiments because it flees reality. Like others in revolt against elevated nineteenth-century sentiments, the medical critics of the
Hippocratic rule assume the stance of tough-minded realists.

Even among laymen the rule Do No Harm is sometimes held
in suspicion, tainted by association with practices of medical paternalism going back to origins of the American Medical Association in the mid-nineteenth century. In 1975 the parents of an
irreversibly comatose patient, Karen Ann Quinlan, sought to
have her removed from the respirator but were overridden by the
hospital and physicians involved, who contended that such an act
would put them in violation of the Hippocratic Oath. Ultimately
the parents prevailed in court, and the case came to represent the
last gasp of paternalistic, that is, Victorian, medicine.

Among the risks of a bias against the past, and particularly the
Victorian past, is that the Hippocratic maxim itself might be sacrificed to it. Thus a critic argues that "The old Hippocratic ethic
saw the patient as a weak, debilitated, childlike victim, incapable
of functioning as a real moral agent," quite as if each and every
patient partook of the helplessness of Karen Ann Quinlan. The
critic concludes with relief that "The Hippocratic ethic is dead."
"Hippocratic" is reduced to an epithet tantamount to "paternalistic," which is to say, "Victorian." Two ethicists highly critical of
the past find dark meaning in the sacralization of the Hippocratic
Oath, beginning in nineteenth-century America. "With the as-
cendancy of scientific medicine, the Hippocratic mandate—'help
the sick *according to my ability and judgment*, but never with a view
to injury or wrongdoing' [emphasis in the original]—has taken on
new force. In the twentieth century, as in the nineteenth, the obligations of beneficence and nonmaleficence have been interpreted and exercised paternalistically." Seen through the lens of a
rigorous hostility to tradition, the clause "according to my ability
and judgment" takes on ominous significance, becoming a warrant for the exercise of domination. But is the clause really so

threatening? Is it wrong for physicians to exercise their ability? Would it be better if they acted like automata? Are they to disregard their knowledge simply because it is theirs? Does aiding another according to one's lights mean trampling that other? If the authors so distrust physicians who exercise their ability and judgment, what would they say about those who *don't* exercise ability and judgment? How can good medicine possibly be practiced except with ability and judgment?

It would be unwise to carry our backlash against the Victorians so far as to reject the principle that guided, in some degree, their practice of medicine. After the best efforts have been made to discredit it, the Hippocratic rule remains an indispensable guide of and constraint on medical practice. A replacement rule like "On the whole do no harm," or "Let the healthcare delivery system in which you are a cog do no harm" or "Do no harm, but remember that mistakes happen" will not do. The principle Do No Harm ought to remain in effect not for reasons of ancestor worship but because medicine dare not lose sight of the inviolability of the patient, and because in our own way we continue to face conditions like those that activated the rule perhaps a century and a half ago. At that time, modern-minded physicians, knowing that folk medicine with its bleeding and purging was of no use and did harm, and possessing more knowledge of disease than power to cure it, came to think of the principle of nonmaleficence as the best guide to practice. It put distance between themselves and the heroic bleeders and purgers (as well as the miscellaneous venders of noxious "cures" who haunt the original AMA Code from one end to the other), and reminded them of the limits of their powers. Medicine still confronts limits that dictate caution—in the finasteride case, limits on medical knowledge especially.

As recently as 1992 the editor of the *British Medical Journal* could report that "only about 15 percent of medical interventions are supported by solid scientific evidence"—such is the "poverty" of medical knowledge. He argued strongly that unless and until the medical profession acknowledges its ignorance, it risks doing harm. Harm *was* done in the case of countless radical mastectomies performed on the basis of a questionable theory and in the face of evidence that it conferred no advantage over more conservative procedures; many a radical prostatectomy may be similarly gratuitous. Or consider the use of oral drugs for hypoglycemia given at one time to patients with adult-onset, non-insulin-dependent diabetes. In 1970 a study of two such drugs had to be halted because both were associated with a significant increase in mortality from cardiovascular disorders, while two more studies shortly thereafter revealed that about half the patients derived no benefit at all from the drugs in question. "One might expect that because of the possibility of harm by agents not required in a large percentage of the patients in whom they had been used, there might be less use of the drugs," and yet their use by physicians, already well established by 1970, continued to grow. Finasteride too poses the possibility of harm to patients with no compelling reason to take it in the first place, but in this case practicing physicians have voted against the drug. Why is this?

Given what is at stake in medicine, the editor of the *British Medical Journal* argues that "our first priority must be to understand the extent of our ignorance." The finasteride papers compel a recognition of ignorance. They force the issue. They put any physician who might think of using the drug for prevention in the position of someone taking a blind leap into a speculative venture, that is, doing something contrary to medical prudence. You cannot read the finasteride papers without being struck by the unsolved riddle of prostate cancer and the poor understanding of this drug whose most ardent advocates concede that the mechanism by which it reduced the incidence of prostate cancer in the PCPT remains unknown. It is true that in and of itself, incomplete understanding may not rule out a given therapy. It has been said that "provided that the Hippocratic injunction to do no harm is not violated, much good can be accomplished during the period between a new therapy's introduction and its scientific validation." (Thus, doctors attending women in childbirth in the mid-nineteenth century could have done much good by disinfecting their hands, as Semmelweis urged, even though the germ theory of disease had not yet taken shape.) As things stand, however, the general use of finasteride would violate that very injunction.

The legendary physician-teacher William Osler, who grew up in the nineteenth century, "was suspicious of what was new and untested and this probably explains his nihilistic [that is, skeptical] approach to therapeutics." Faced with the paradox of finasteride, medicine has in effect gone back to its Victorian sources for counsel. It views the drug much as Osler might have, that is, skeptically, concluding that even after a carefully designed clinical trial which involved hundreds of centers, thousands of volunteers, millions of dollars, and years of meticulous effort, the value of finasteride remains unknown and in doubt.

Osler saw the nineteenth century just as, according to Walter Houghton, the Victorians of England saw their age: as a time of transition. Looking back over the remarkable course of medical progress in the nineteenth century, Osler was particularly struck by the far more discriminating use of drugs instituted over that period. The beginning of the century saw all manner of drugs—noxious and nauseating on the one hand, innocuous but inactive on the other—used in all manner of cases. What Osler designates the New School of medicine confined itself to the use of a few drugs of proven worth.

> The battle against poly-pharmacy [he wrote in 1901], or the use of a large number of drugs (of the action of which we know little, yet we put them into bodies of the action of which we know less), has not yet been fought to a finish. . . . [The New Medicine] is more concerned that a physician shall know how to apply the few great medicines which all have to use, such as quinine, iron, mercury, iodide of potassium, opium and digitalis, than that

he should employ a multiplicity of remedies the action of which is extremely doubtful.

Applied to the finasteride question, Osler's medical skepticism seems to say: "Remember how little is known about this powerful drug and its effect on carcinogenesis. Remember that drugs must be used discriminately, with due caution, and that it was only when medicine began to do so that it entered the age of progress." And for now, this advice has been heeded.

3. Roots in the Past

Among the institutions that participated in the PCPT is a center of prostate research where I have received treatment: the University of California, San Francisco (guarded by a statue of Hippocrates). Situated as it is in a city proud of its sexual liberation but rich with late-Victorian architecture, this university symbolizes for me medicine's roots in a discredited past.

Even the concern that only a fraction of medical practices is "supported by solid scientific evidence" has roots in the past, for it was in the Victorian era that the question of the evidentiary basis of medical treatment became both salient and acute. The old way of purging and more purging, as if the answer to failure were more of the same, had begun to creak and strain under its own absurdity. Consider a case that took place in Massachusetts in 1809, involving the founder of the medical sect known as the Thomsonians, which was still going at midcentury. Thomson

> was called to the home of Ezra Lovett, a young man who had been confined to his bed for several days with a severe cold. Thomson wrapped Lovett in hot blankets and gave him dose after dose of powerful emetics, which induced violent vomiting. . . . Thomson prescribed dozens of emetic and sweating procedures over several days. When the emetics ceased to work,

Thomson administered a cathartic (laxative) concoction, and Lovett went into convulsions—at which point Thomson's assistants held the thrashing Lovett so that Thomson could administer yet another emetic dose. After three days of almost constant convulsions, Lovett died.

So wedded to his theories was Thomson that he was willing to kill the patient in their name, as indifferent to his own wrongdoing as he was to the very concept of evidence. The nineteenth-century revolution in medical theory and practice required a rethinking of the nature of evidence, and in this project John Stuart Mill had a part.

Deep in the middle of the work that made his name in his own time—his *System of Logic*—Mill investigates the difficulty of establishing that a given drug such as mercury cures a given disease. He concludes that because of the ineliminable presence of confounding factors, no experiment could prove the efficacy or inefficacy of the drug.

> The mercury of our experiment being tried with an unknown multitude (or even let it be a known multitude) of other influencing circumstances, the mere fact of their being influencing circumstances implies that they disguise the effect of the mercury, and preclude us from knowing whether it has any effect or no. Unless we already knew what and how much is owing to every other circumstance (that is, unless we suppose the very problem solved which we are considering the means of solving), we cannot tell that those other circumstances may not have produced the whole of the effect, independently or in spite of the mercury.

Here Mill appears to deny the possibility of the sort of clinical trial that has since become the standard of medical verification. In this he was wrong—but no matter. His ethos of scrupulous analysis and his high standards of evidence helped lay the basis for trials, such as the PCPT, in which these principles are observed.

Mill's question, "Does or does not mercury tend to cure [a] particular disease?" feeds eventually, like a tributary, into the question, "Does finasteride prevent prostate cancer?"

That Mill was wrong to dismiss the possibility of experimental verification of a drug's effect doesn't mean that his conclusion was without practical import. If so little is known about a drug that the most carefully contrived experiment will not establish its effect with any certainty, it follows inevitably, morally speaking, that those in the business of treating disease must proceed cautiously, with a due wariness of causing harm. In effect, the principle of philosophical modesty and the principled use of drugs become one.

[11]

The Finasteride Story:
What Did Not Happen

1. COOL ARDOR

Despite the disrepute of tradition these days, the inherited principle, Do No Harm has kept medicine from embracing finasteride, a drug some believe would be worth using on a grand scale
even if its risks were considerably worse than they now appear.
Following medicine's lead, advocacy groups, the press, and patients themselves have all decided against the chemopreventive
use of finasteride, for now. This general agreement to err on the
side of caution is highly fortunate, for without it we might be
led into a serious error by the ardor of our own hopes, the desire
for magic bullets, aggressive rhetorical marketing, slick ads, the
engaged passions of researchers, the investment of funds and
effort in such an endeavor as the PCPT, or sheer intolerance of
uncertainty.

As the original report on the PCPT was about to be published
in the *New England Journal of Medicine* in 2003, one of the study's
directors declared in a news release, "I am firmly convinced that
this is the first step in conquering prostate cancer." The evangelical ring of this proclamation of faith, so contrary to the spirit of

both dispassionate inquiry and careful practice, is nowhere to be found in the actual medical literature on finasteride. It belongs to the world of publicity, bringing to mind the exclamatory journalism of M. Homais, correspondent of the Rouen *Beacon*. On the one hand a clinical trial that augurs the conquest of prostate cancer, on the other "a great surgical experiment" promising not only the cure of a clubfooted patient but the conquest of superstition itself ("What fanaticism once promised to its elect, science shall now accomplish for all mankind!"), the tone in both cases being visionary and ecstatic. Homais composes, on the side, a *General Statistical Survey* of his district. The 2003 news release quotes the researcher as saying that, statistically speaking, the benefits of finasteride far exceed the risks, if any.

The headline of the news release reads, "[University of Texas at San Antonio] Health Science Center Urologist in Lead as U.S. Announces Finasteride Reduces Risk of Prostate Cancer by 25 Percent." With these trumpet notes still in the air, the story acknowledges the increased risk of high-grade tumors on the finasteride side of the PCPT, but not until the seventh of nine paragraphs is the issue really confronted.

> Although fewer men on finasteride developed cancer, those who did ran a greater risk of having a more-serious cancer than the men who developed cancer in the placebo group. "We don't know if the increase in higher-grade tumors was 'real' or not," Dr. Thompson said. "Previous studies have suggested that tumors diagnosed in men on hormonal therapy appear to be higher grade when they really are not. If we assume, however, that the increase was correct, the magnitude of the increase in higher-grade tumors was still much less than the reduction in overall cancers."

Evidently the researcher is firmly convinced that finasteride heralds the conquest of prostate cancer not because he is sure the drug poses no serious dangers but because he is sure that even if those dangers exist, they are statistically overshadowed by benefits. The significance of high-grade tumors is dismissed with a wave of the wand.

Aside from a few investigators (of whom Dr. Thompson is one) persuaded that even if finasteride's risks were twice those in the PCPT it would still make sense to use on a general scale, few in the medical literature seem to think the danger of high-grade cancer sinks into triviality next to the drug's benefits. Precisely because they take such a danger seriously, many of the finasteride papers investigate the possibility that it is not real. Medical papers are in any case not written in the tone of a sales convention, a pep rally, an annual report, or a feature in the Rouen *Beacon*, nor do their authors speak as if the imperative Do No Harm belonged to the dark prehistory of medicine.

Indeed, in his professional papers Dr. Thompson himself—the same man who foretold the conquest of prostate cancer—often writes cautiously, with a due respect for the unknown. In print Dr. Thompson does not proclaim. He does not write as if the scroll of the future lay open before him, and does not declare his faith in finasteride but evaluates the evidence in its favor, aware of the counterevidence. He does not, as a rule, imply that the risk of high-grade cancer just doesn't signify. (This is especially true of an article, distinguished by its measured language and qualified conclusions, co-written by Dr. Thompson for the *British Medical Journal*. Seemingly adapted to the more skeptical ethos of prostate cancer medicine in the United Kingdom, the paper plays down finasteride, simply noting at one point that "one randomized trial showed that finasteride, a 5α reductase inhibitor, reduced the risk of incident prostate cancer at seven years by about six men in 100. Detected cancers considered high grade, and thus potentially of greater risk for causing morbidity and mortality, were more common in the finasteride than the placebo group." Even so, the paper concludes that "finasteride reduces the risk of prostate cancer.") Some may say that a researcher will naturally weigh his words in medical journals where caution is the rule, and express himself more exuberantly in press releases where subtleties and fine points are out of place and rules relaxed. But it is not true that the press is incapable of grasping the ambiguity of the finasteride data, for reports on the PCPT in 2003 did just that (as did reports on the BCPT in 1998). The *Wall Street Journal* story on the PCPT on June 24, 2003, began, "Medical researchers have found a drug that sharply reduces the risk of prostate cancer. But their findings present a painful dilemma for men because taking it may cause worse tumors in an unlucky few." Of course if finasteride were used "at the population level," as its defenders and enthusiasts envision, the unlucky few would grow by orders of magnitude. And the drug's partisans know this. To me the stark disparity between Dr. Thompson's proclamation of faith in finasteride and his measured conclusions as a writer of medical papers suggests that a certain zeal is being held in check—beneficially—by the norms of medicine.

The original *New England Journal of Medicine* report on the PCPT, of which Dr. Thompson is the principal author, frankly concludes that "finasteride prevents or delays the appearance of prostate cancer, but this *possible* benefit . . . must be weighed against sexual side effects and the increased risk of high-grade prostate cancer" (my emphasis). Findings are couched in an uncertainty appropriate to a disease whose biology remains "unpredictable." "It is possible that finasteride induces high-grade tumors. . . . Long-term follow-up . . . and further laboratory research will be required to determine the reason for the association between finasteride and high-grade prostate cancer." The higher rate of positive biopsies among men in the finasteride group "could" result from detection bias. The doubts and dangers played down in the press release are thus given their full weight in the pages of the *New England Journal of Medicine*; nor does the *NEJM* paper quite state that if we subtract cost (an increase in high-Gleason tumors) from benefit (cases of cancer spared), the result is a summary judgment in finasteride's favor. The question of harm casts a passing shadow over the press release but a deep shadow over the paper.

Like a person restrained by his better judgment, some papers in the medical literature review the benefits of finasteride and parry the arguments against it only to admit in the end that the risks standing in the way of the drug's general use may well be real. Such a paper will note that the elevated rate of the most dangerous tumors in the finasteride group of the PCPT flattened out after the first year (which, some say, argues against the likelihood that finasteride actually contributes to cancer); that finasteride reduces the volume of the prostate, making it more likely for biopsy needles to hit cancer; that finasteride makes PSA a more effective cancer detector (which not only helps explain a higher detection rate in the finasteride side of the PCPT but argues for the use of finasteride for screening purposes); that because finasteride inhibits low-grade disease, it seems, but only seems, to favor high-grade disease; that examination of surgically removed prostate glands in the PCPT reveals that detection of high-Gleason disease was more likely in the finasteride group (meaning that such disease was more often missed in the other group); and that the PCPT may actually have underreported the preventive effect of finasteride. Having assembled this circumstantial case for finasteride, the paper will then conscientiously concede that the drug may after all induce aggressive cancers and cannot therefore be recommended for the purpose for which the PCPT was conducted: prevention. This is not a case of a mountain giving birth to a mouse but of research advanced and informed by the tradition of intellectual modesty.

Similarly, a recent review of the PCPT, after surveying several possible reasons for the increased rate of high-grade malignances except that finasteride might actually have caused them, nevertheless grants in the end that "concerns regarding the higher prevalence of high-grade disease associated with finasteride therapy" have ruled out the drug's "widespread use." Given the demonstrated preventive effect of finasteride, given a number of invitingly plausible explanations of the more ominous PCPT data, given the massive investment of effort in the PCPT itself (in which one of the authors of the review in question played a large part), the paper might well have been expected to endorse finasteride for general use. It does not.

Two weeks after *Business Week* declared the PCPT "a big win for . . . chemoprevention," it qualified this simplistic judgment with the observation that "the men on Proscar [in the PCPT] who did develop cancer tended to get more deadly tumors, possibly because of the drug." Such course corrections and expressions of reduced enthusiasm were not confined to the press. After confessing that the reduced rate of cancer in the finasteride group "blew me away," a director of the PCPT acknowledged that the study yielded "a mixed message." A headline in the medical press dramatizes the sort of recklessness that might follow if the tradition of cautious inference and practice were abandoned. While the article in question concludes its summary of a follow-up paper on the PCPT by noting that "The increased risk of high-grade disease in the PCPT was due, *at least in part*, to improved detection, rather than the induction of high-grade disease" (my emphasis), the headline reads, "Proscar Exonerated as Trigger for High-Grade Prostate Cancer." Somehow the finding that some degree of detection bias was at work in the PCPT is inflated rhetorically into a full pardon for finasteride.

In a recent story on Propecia, which is Proscar at a lower dose (the original press reports of the PCPT confused the two), Reuters noted that men in the finasteride arm of the PCPT ran a higher risk of aggressive cancer but, according to a researcher, probably not because of the drug. "The reason that probably happened," he stated, is that the study's designers did not adequately compensate for finasteride's effect on PSA. Sometimes the finasteride controversy reads like a debate between hope and caution, with neither party silenced for long. Following the PCPT, Dr. Peter Scardino of the Department of Urology at Sloan-Kettering Cancer Center cautioned against using finasteride for prevention, concluding that "on balance, finasteride does not seem to be an attractive agent for the chemoprevention of prostate cancer" because it may suppress cancers not worth treating only to induce genuinely dangerous ones. Later Dr. Scardino revised his view, concluding that finasteride "may be safe and effective in preventing prostate cancer," even proclaiming that "chemoprevention of prostate cancer is at hand." At a recent symposium on prostate cancer, however, Dr. Scardino acknowledged

> that a higher standard needs to be used before advocating wide scale medication for prevention, as even small or rare effects will reach a lot of people. He claimed the cancers prevented by Finasteride were ones which would never have needed treatment anyway.

Only because of the well-established principle of erring on the side of caution does the burden of proof in this debate fall squarely on those in favor of the general use of a drug whose safety is in question. Their arguments must meet "a higher standard" of evidence, and to this point they have not done so.

2. WHAT DID NOT HAPPEN

Running through the finasteride papers as well as the press response to the PCPT, then, is a current of restraint—an influence that warns against a hasty embrace of the drug, cautions against playing down evidence of its risks, reminds us of its unknown effects, puts us on guard against the fervor of our own hopes. If not for this admonitory influence, both the medical world and the world at large might have rushed to acclaim a pill that could markedly reduce the incidence of a cancer both common and dreaded. As for the risks of this new wonder drug, they could be written off as incidental casualties of progress, or left to look after themselves, or explained away as more apparent than real. The admonitory force at work in the finasteride debate is the prohibition against doing harm in the name of medicine—a prohibition dating to the beginnings of the medical guild itself, over its history honored more often in the breach than the observance, revived by the Victorians, both violated and even deplored in our time, but still possessing the force of an ideal and a norm.

The deterrent effect of the Hippocratic principle is strongly in evidence in the very abstract of a 2005 paper on "The Effects of 5 α-Reductase Inhibitors on the Natural History, Detection, and Grading of Prostate Cancer." With the effect of gunning an engine with a car in neutral, the authors cite one merit of finasteride after another, only to conclude that the drug cannot yet be recommended for prevention after all. I cite this passage, translating it point by point into common language to bring out the braking effect at work:

> Results: In PCPT there were fewer biopsies performed for cause in the finasteride vs the placebo group and the proportion of high grade tumors did not diverge with time. [*Implication: finasteride does not cause high-grade tumors.*] . . . Prostate shrinkage in the finasteride treated group may minimize biopsy sampling error. [*Implication: by shrinking the prostate, finasteride simply made it easier to detect cancer; it did not cause cancer.*] Furthermore, histological changes have shown that 5ARIs have a significant effect on prostate architecture, which can make the interpretation of prostate specimens treated with 5ARIs difficult. [*Implication: the Gleason scores assigned to finasteride-treated cells are unreliable because of the cellular changes produced by the drug.*] Further evaluation of PCPT findings will help determine the true nature of these observations.
>
> Conclusions: 5ARIs decrease the risk of prostate cancer but also alter the detection of disease through effects on prostate spe-

cific antigen, and prostate volume and histology. [*Implication: not just one but three detection biases worked against finasteride.*] The weight of evidence suggests an artifactual effect of finasteride on Gleason grading in the PCPT. [*Implication: the increased rate of high-grade cancer in the treated group is most likely a mirage.*] The role of 5ARIs for prostate cancer chemoprevention needs further examination before it can be considered for wide recommendation.

The final sentence, recommending *against* the preventive use of finasteride at this moment, hangs in space like a non sequitur, contradicting the effect of all that precedes it. Aside from the force of caution—that is, a concern over doing harm—what else could possibly account for a conclusion so plainly at variance with every one of the authors' inferences and arguments? And the same restraint is at work throughout the finasteride story, both in the press and in medical papers and practice.

As already noted, at the start of the tamoxifen trial a *New York Times* writer predicted that "If the five-year study shows that the hormone prevents cancer, millions of women will probably be advised to take it for the rest of their lives." Although tamoxifen was shown to prevent breast cancer in the BCPT, the millions of American women in the same risk category as the volunteers in the BCPT were not given (and did not demand) the drug in the expectation that breast cancer among them would decline by 50 percent and that the side effects of the drug, such as endometrial cancer and stroke, would take care of themselves. This sort of medical carpet bombing did not take place. As with tamoxifen, the finasteride story is in good part a story of what did not happen.

Consider then what did not happen following the PCPT:

§The media did not shower confetti on the results of the PCPT.

§Despite its activist approach to prostate cancer and its promotion of early detection, American urology did not take the next

step and recommend the use of an effective chemopreventive—
even though the revised Hippocratic Oath pledges the initiate to
"prevent disease wherever I can, for prevention is preferable to
cure."

§Despite circumstantial evidence in finasteride's favor, despite
the possibility of several biases conspiring to inflate the number
and grade of malignancies charged to the drug in the PCPT, despite their intuition that the drug "almost certainly" does not fuel
aggressive cancer, physicians still did not prescribe it to prevent a
disease that badly needs prevention.

§The medical world did not agree that unless the rate of high-
grade cancer tripled among men taking finasteride, "The potential detrimental effects of an increased rate of patients who have
prostate cancer with high-grade Gleason scores would be outweighed by a reduction in incidence." Many, I imagine, were unable to contemplate without concern a simple doubling of the
rate of high-grade cancer. Some were troubled by the merely elevated incidence of such cancer in the PCPT. The medical world
did not conclude that because the number of cancers prevented
by finasteride in the PCPT exceeded the number of high-grade
cancers induced, the drug yields a net benefit.

§Urologists did not say, "We are already prescribing finasteride; why should we not take advantage of its preventive effect
by prescribing it to more men at the same dosage?"

§The medical profession did not take the position that the
principle Do No Harm represents so much "pointless nostalgia,"
an imposition of the past on the present, an archaism that does
not "tally any more with the realities of modern society," an ob-
stacle to the pursuit of the greatest good for the greatest number.
It did not argue that the Hippocratic maxim is bound up with ancient medical theories and should have died when they died.

§Despite the general depreciation of tradition in postmodern
society, writers on the finasteride question did not cast off the tradition of philosophical modesty. Despite the antiquity of the Hip-

pocratic principle, despite its Victorian associations, despite the lying that was done in its name, doctors confronting the finasteride question were not prepared to discard it, either, as a remnant of the past. Despite the astounding advances of medicine since the days of Jefferson, they still thought it necessary to refrain from "conjectural experiments on a machine so complicated and so unknown as the human body, and a subject so sacred as a human life."

Afterword

AN INTERVIEW WITH MY UROLOGIST

SJ: Do you prescribe finasteride for preventive purposes?

MD: No.

SJ: Do you know others who prescribe it?

MD: Not routinely. My partner has a strong family history [of prostate cancer], and I know he's choosing not to use finasteride himself.

SJ: In spite of being at being at elevated risk, he chooses not to take finasteride?

MD: Right.

SJ: Some say the higher numbers of high-grade malignancies associated with finasteride were an artifact of the PCPT itself. Do you think the risk is real?

MD: The data were real, but there is no known biological mechanism by which finasteride would raise the grade of tumors. Most urologists would suspect some sort of detection bias at work. After all, if finasteride suppresses cancer, why would it increase high-grade cancer?

SJ: You suspect the risks of finasteride aren't real. What would it take to convince you they aren't real?

MD: More data. I read the original *New England Journal of Medicine* article and believe some additional analysis is due to appear.

SJ: Do you prescribe finasteride for BPH?

MD: Infrequently.

SJ: How do you justify using the same drug in one case that you wouldn't use in another?

MD: First, I don't prescribe it that often—maybe twice a year. The costs and side effects of the drug are considerations, as is the waiting time for results [in cases of BPH]. There are better alternatives out there. But prevention means giving the drug indiscriminately to anyone with an XY chromosome as opposed to people with a known indication.

SJ: Has Merck sent you samples of Proscar?

MD: I think they used to.

SJ: Do you see yourself changing your views on the use of finasteride?

MD: I'd consider using it for people with risk factors.

SJ: Family history?

MD: Right. If my dad died in his fifties [of prostate cancer] and I had a couple of uncles who also died of the disease. But would I recommend that your son take finasteride preventively at some point? No. He should be seen and checked regularly, no doubt about that. But finasteride—I'd say no.

SJ: What do you think of the researchers who really pitch finasteride?

MD: I saw one of them at a convention once. Was he himself taking finasteride? I can't remember. But I do remember, also at a urological convention, a speaker asking, "Who here takes a vitamin?" and all around the room hands shot up. Then he asked, "Who here takes finasteride," and there was not a single hand. And this at a gathering of urologists.

SJ: I know one researcher who takes it as a hair drug. Propecia.

MD: Not at the five-milligram dosage, though.

SJ: Why do you think deaths from prostate cancer have remained constant in spite of the screening revolution?

MD: Have they remained constant? I think there may have been some decrease in recent years.

SJ: Okay, why have they remained relatively constant while the number of diagnoses of prostate cancer has gone skyward?

MD: I think because of the natural history of the disease.

SJ: Do you mean we're not catching it early enough?

MD: We're not catching it early enough *in those who are dying from it.* Remember, we screen for PSA, and there is no PSA however low that guarantees the absence of prostate cancer. So in some cases by the time PSA screening catches up to cancer it is already advanced. The screening techniques in use today— a finger and PSA—provide only crude measures. I put more hope in improved screening techniques than in finasteride.

SJ: In the PCPT, at the end of the trial, men who showed no warning signs of cancer—no abnormal DRE or elevated PSA—voluntarily underwent biopsies, and a shocking percentage of were found to have cancer.

MD: Yes. Their low PSAs did not prove the absence of cancer. Some men like these have prostate cancer that eludes PSA screening, and by the time it is caught it is already late in the day. That I think is why the death rate has remained what it is, even in the PSA era.

SJ: What if one of your patients asked for finasteride?

MD: I would consider it. I wouldn't say no. But would insurance pay for it?

SJ: Do you think finasteride is the beginning of a prevention revolution?

MD: I don't think so. In any case I would need more data before I would recommend that everybody take finasteride preventively. That's just too universal, too indiscriminate.

Notes

1. A Medical Maze: The Finasteride Question

page

3 "conquerable by human care and effort": "Utilitarianism" in *Essential Works of John Stuart Mill*, ed. Max Lerner (New York: Bantam, 1971), p. 202.

4 "that the tumors have been permanently eliminated": see paper 23 (as numbered in the Bibliography).

4 *"reduce the risk* of breast cancer, not to prevent it." Michael W. DeGregorio and Valerie J. Wiebe, *Tamoxifen and Breast Cancer* (New Haven: Yale University Press, 1999), p. viii.

4 prostate cancer "prevention" in this qualified sense of the word. On the possibility that the preventive effect of finasteride may be purely temporary, see paper 14.

4 a clinical trial designed to yield probative evidence. On clinical trials and chemoprevention, see paper 40.

5 "the first step in conquering prostate cancer": University of Texas Health Science Center at San Antonio news release, June 26, 2003.

5 "tempered enthusiasm": see paper 25.

5 "finasteride's ability to reduce prostate cancer risk": National Cancer Institute news release, June 24, 2003.

5 "a very big day for men's health": CNN, June 24, 2003.

7 "no genotoxic or other adverse properties": see paper 7.

7 "enhance rather than slow down the progression of prostatic cancer": see paper 6. It is Dr. Bosland's view that "there is really no future for finasteride as a preventive agent against prostate cancer unless continued follow-up of this study shows real benefits (better survival, etc.)." Personal communication.

8 a promise not to harm. Louis Lasagna's 1964 version of the Hippocratic Oath appears at www.pbs.org/wgbh/nova/doctors/oath_modern.html.

10 even as medicine modernized: Much as shadows of the heroic haunted the more modern, Lockean literary form of the novel.

10 "irreparable errors and wrongs": Oliver Wendell Holmes, *Medical Essays, 1842–1882* (Boston: Houghton, Mifflin, 1895), p. 167. Holmes's pamphlet on puerperal fever was published in 1843.

10 "the multiform and often fatal injuries": Worthington Hooker, *Physician and Patient* (New York: Baker and Scribner, 1849), p. 25.

11 "when the first of the wonder drugs was introduced": Edward Shorter in *Cambridge Illustrated History of Medicine*, ed. Roy Porter (Cambridge, UK: Cambridge University Press, 1996), p. 128. Over some forty years in the nineteenth century, four diseases were identified that still bear the names of those who first described them: Bright's, Parkinson's, Hodgkin's, and Addison's. In each case, while physicians could not cure, at least they now possessed the knowledge to refrain from harming the patient with some senseless treatment. On the avoidance of harm, see Anne Hardy, *Health and Medicine in Britain Since 1860* (New York: Palgrave, 2001), p. 27: "Good medical management increasingly helped to prolong life after 1860 when patients had died more quickly in earlier decades."

12 sincerely believed the truth can cause harm. Thus, for example, Oliver Wendell Holmes maintained that "Your patient has no more right to all the truth you know than he has to all the medicine in your saddle-bags. . . . He should get only just so much as is good for him. . . . It is a terrible thing to take away hope, even earthly hope, from a fellow-creature. Be very careful what names you let fall before your patient." Holmes, *Medical Essays*, p. 389. Even so, doctors' reluctance to destroy hope did not necessarily lead to lying. The original (1847) Code of Medical Ethics of the American Medical Association recommends not that doctors withhold bad news from the patient but that the task of informing the patient and his or her dear ones "be assigned to any other person of sufficient judgment and delicacy" (Chapter 1, Article 1, Section 4).

12 could have been discarded some decades ago when the entire edifice of paternalism fell into disrepute. The original proponents of anesthetizing women in labor insisted that chloroform was perfectly safe if carefully administered. In taking childbirth off women's hands, these men no doubt acted paternalistically. Paternalism and the duty not to harm thus went hand in hand in this setting. It is significant that the revolt against medical paternalism (in the name of autonomy or self-determination) coincided with the "natural childbirth" movement.

12 "obligation 'to be of benefit and do no harm'": Albert R. Jonsen, *The New Medicine and the Old Ethics* (Cambridge, Mass.: Harvard University Press, 1990), p. 67. Osler described the Hippocratic Oath as "the high-water mark of professional morality. . . . For twenty-five centuries it has been the 'credo' of the profession." William Osler, *The Evolution of Modern*

Medicine (Kila, Mont: Kessinger, n.d.), p. 46. Osler's lectures on the history of medicine were originally delivered in 1913.

12 "the fatal termination of many illnesses was attributable to the doctor rather than to the disease": A. J. Youngson, *The Scientific Revolution in Victorian Medicine* (New York: Holmes and Meier, 1979), p. 18.

13 "simply did not work": Lewis Thomas, "Biostatistics in Medicine," *Science*, 18 November 1977.

13 "possible remedies for disease much more powerful and potentially dangerous": ibid.

13 as in the case of finasteride. Skepticism is hardly obsolete. According to the seminal work on evidence-based medicine, A. L. Cochrane's *Effectiveness and Efficiency: Random Reflections on Health Services* (London: Royal Society of Medicine Press, 1972; rpt. 1999), "environmental factors alone were important in improving vital statistics up to the end of the nineteenth century and . . . until the second quarter of this century therapy had very little effect on morbidity and mortality. One should, therefore, forty years later, be delightfully surprised when any treatment at all is effective, and always assume that a treatment is ineffective unless there is evidence to the contrary" (p. 8).

13 There is no denying that folk practices . . . harmed. On the other hand, the use of medicinal plants could help. Recent research on such plants has shown that "many of those included in the herbal pharmacopoeias of Eastern and Western medicine were genuinely effective. Some, like quinine, quickly became famous." Alan Macfarlane and Iris Macfarlane, *The Empire of Tea: The Remarkable History of the Plant That Took Over the World* (Woodstock, N.Y.: Overlook, 2003), p. 282. Among the most traditionally esteemed of medicinal plants is the tea plant, celebrated for its curative properties in both East and West. Tea may in fact be of medicinal value; what is not in doubt is that the preparation of tea with boiling water, by killing waterborne agents of disease and thus making water safe to drink, proved one of the most momentous health benefits in human history.

14 "has necessarily interfered with the Hippocratic ideal of medicine": see paper 43.

14 "villagers cluster around a healer and a patient": Jonsen, *New Medicine*, p. 1.

14 not the center of the medical universe. Some, however, would say that patients have only just been placed "at the center of medical decision-making." Virginia Sharpe and Alan Faden, *Medical Harm: Historical, Conceptual, and Ethical Dimensions of Iatrogenic Illness* (Cambridge, UK: Cambridge University Press, 1998), p. 116.

14 "the benefits from . . . tamoxifen therapy must be weighed against the risks and costs": DeGregorio and Wiebe, *Tamoxifen and Breast Cancer*, p. 39.

15 "population based rather than patient based": Jonsen, *New Medicine*, p. 30.

16 "simple bed rest predisposes to thromboembolic disease": see paper 56. Measures can be taken against these things, however. It is not as if they were inevitable.

17 congressional hearings were held. John R. Paul, M.D., *A History of Poliomyelitis* (New Haven: Yale University Press, 1971), p. 437.

17 "the positive effects would diminish, or even disappear": H. Gilbert Welch, *Should I Be Tested for Cancer? Maybe Not and Here's Why* (Berkeley: University of California Press, 2004), p. 167.

17 actuaries, statisticians, or investors. Ironically, "somewhere around the age 50" prevention ceases to make much economic sense, as it then merely prolongs a diminishingly productive life. See Geoffrey Rose, *The Strategy of Preventive Medicine* (Oxford: Oxford University Press, 1992), p. 3. Enrollees in the PCPT were at least fifty-five.

18 "each time the data are discussed": see paper 52.

19 "certainly was recognized in Germany": ibid.

19 "does not mention informed consent and disavows surgery": Steven Miles, *The Hippocratic Oath and the Ethics of Medicine* (Oxford: Oxford University Press, 2004), p. 2.

19 or that placebos do no harm. See an anecdote recorded by Hooker, *Physician and Patient*, p. 82: "An amusing instance of the celebrity sometimes gained by inert remedies, occurred in Paris. A man who had sold to great profit an eye-water, at length died without communicating to any one the composition of it. His widow regretted the loss of the profits. . . . Without telling her trouble to any one, she filled up the phials from the River Seine, and went on to sell the eye-water as usual. Cures occurred as before, and everybody believed that her husband had bequeathed the recipe to her. On her death-bed her conscience was much disturbed on account of the deception which she had thus practiced upon the community, and she made confession to the physician who attended upon her. He, however, quieted her mind by telling her, that he was sure she need give herself no uneasiness, for her medicine had at least done no harm—a consolation which most venders of secret medicines could not have."

19 "the physician's hand or fingers were to be cut off": Sharpe and Faden, *Medical Harm*, p. 7.

20 "often enough with no medicine at all": Sydenham citing Hippocrates; in Osler, *Evolution of Modern Medicine*, p. 126.

21 "the moment at which doctors began to be able to save lives": David Wootton, *Bad Medicine: Doctors Doing Harm Since Hippocrates* (Oxford: Oxford University Press, 2006), p. 5.

21 "almost all ethical doctrines and religious creeds . . . a few phrases retained by rote": John Stuart Mill, *On Liberty* (New York: Norton, 1975), p. 39.

22 "undeceived consent and participation": Mill, *On Liberty*, p. 13.

23 "that it would increase the average life by some days": D'Alembert's "Memoir" on smallpox inoculation, included in L. Bradley, ed., *Smallpox Inoculation: An Eighteenth Century Mathematical Controversy* (Nottingham: University of Nottingham, 1971), p. 63.

24 a condition then treated with surgery. It was said at the time that "prostatectomies for the relief of prostate enlargement are one of the most common surgical procedures in the United States," more than 300,000 being performed annually. "When the surgically removed tissue is examined in the lab, 10 per cent of the men—potentially over 30,000 a year—are discovered to have some traces of cancer." Louise Russell, *Educated Guesses: Making Policy about Medical Screening Tests* (Berkeley: University of California Press, 1994), p. 39.

24 Finasteride was first given experimentally to human subjects in 1986. See papers 53 and 59.

2. The Prostate Cancer Prevention Trial

25 "would have lived as long without being a cancer patient": H. Gilbert Welch, Lisa Schwartz, and Steven Woloshin, "What's Making Us Sick Is an Epidemic of Diagnoses," *New York Times*, January 2, 2007.

25 people lived no more than a year or so once cancer became apparent. See paper 33.

26 fully half the cases of discovered disease. See paper 74.

26 "some form of definitive therapy instead of watchful waiting": see paper 64.

26 "no cancer in which the existence of pseudodisease is more widely accepted and the value of early detection more vigorously questioned": Welch, *Should I Be Tested for Cancer?*, p. 63; see also paper 10.

27 "would be difficult to find": see paper 76.

27 It was to test this possibility that the Prostate Cancer Prevention Trial (PCPT) was conducted. See paper 67. In progress is an international clinical trial of another, and possibly more effective, 5 alpha-reductase inhibitor, dutasteride. Known by the acronym REDUCE, for Reduction by Dutasteride of Prostate Cancer Events, the trial will study men at *high* risk of prostate cancer.

27 at comparatively low risk for prostate cancer. Like men at high risk, men with a history of severe BPH were excluded from the PCPT, which explains the low rate of BPH in the PCPT data.

28 "depends on how hard you look": Welch, *Should I Be Tested for Cancer?*, p. 59.

29 "multiple effects, both positive and negative": see paper 40.

29 has not made much of a dent in prostate cancer's death rate. Note, however, that in the UK, where PSA screening is viewed more skeptically and

proportionally far fewer men are diagnosed with prostate cancer, proportionally far more die of it—9,500 annually.

30 "but a higher risk of high-grade cancer": see paper 65.

30 it would look as if the second offset or overshadowed the first. Depending on how it is calculated, the increase in high-grade tumors would be as high as 68 percent. See papers 38 and 25. In the BCPT, the rate of breast cancer was reduced 49 percent, but the rate of endometrial cancer increased 136 percent; absolute numbers were lower in the case of endometrial cancer, though.

30 than men with the lowest grades have of dying within twenty years. See paper 74.

30 "even one slide that shows a high-grade lesion or cancer": cited in Russell, *Educated Guesses*, p. 80.

30 "a third of deaths from prostate cancer occur in men with Gleason 6 or less disease": see paper 64.

32 "unnecessarily and imprudently treating asymptomatic men with BPH": see paper 34.

32 now with its safety in question? As I write, the Proscar web page itself prominently reports the rates of high-grade cancer on both sides of the PCPT, adding that "the clinical significance of these findings is unknown" and that "Proscar is not approved to reduce the risk of developing prostate cancer." As if to avoid conveying the impression that the drug does just that, the web page does not even mention the reduced rate of prostate cancer in the finasteride arm of the PCPT—a significant act of abstention on the part of the company, Merck, that synthesized finasteride and stands to gain from its use on a grand scale. Note, however, that Merck's patent on finasteride at the five milligram dosage has expired. Generic finasteride will soon be available.

32 "was deeply entrenched in obstetrics practice": see paper 5.

33 "should not be undertaken until further studies have led to a better understanding of such drugs": cited in Richard Gillam and Barton Bernstein, "Doing Harm: The DES Tragedy and Modern American Medicine," *The Public Historian* 9 (1987): 64.

34 "carcinogenic effects that may take years to manifest": see paper 5.

34 it was responsible for tens of thousands of heart attacks: see paper 35.

34 "must . . . be very safe because of the very long duration of their application in men at risk of prostate cancer": see paper 7.

35 "men [in the PCPT] diagnosed with prostate cancer prior to the end-of-study biopsy saw a benefit from finasteride": see paper 19.

35 "must be weighed against the smaller absolute increase in the risk of high-grade disease": see paper 67.

35 as if finasteride lowered the cancer risk overall. As also in paper 66.

37 pales into insignificance next to so great a number as 4.6. See paper 36.

37 "side effects would have to be minimal in order to achieve an acceptable ratio of benefit to risk": see abstract of R. C. Bast, et al., "Prevention and Early Detection of Ovarian Cancer: Mission Impossible?" *Recent Results in Cancer Research* 174 (2007).

37 "which do not at all, or only to a minimum extent, attack or damage the organs of the body": Ehrlich as quoted by Martha Marquardt, *Paul Ehrlich* (London: Heinemann Medical Books, 1949), p. 91. Ehrlich took reports of adverse reactions to Salvarsan very seriously. In one case a Dr. Almqvist of Stockholm visited Ehrlich with word that among many patients successfully treated, one had died. Dr. Almqvist later wrote, "It was my first case of death after Salvarsan treatment, caused by encephalitis haemorrhagica interna. I had to tell him [Ehrlich] every detail of the treatment, the way in which the disease became progressively worse, the result of the post-mortem examination. Although continually interrupted by visitors he always returned to the discussion of this particular case, walking up and down his room, at one minute asking animatedly about the details, the next silently meditating. . . ." (p. 217).

38 "could outweigh any benefit from an overall reduction in cancer": see paper 81.

39 "This effect could introduce a bias against any evidence of benefit from finasteride": see paper 67.

39 recommending changes in biopsy technique. See paper 27.

40 "may simply result from prostate volume reduction": see paper 38.

40 "where the probability of sampling a nondominant cancer is higher": see paper 38.

40 an accurate measure of cancer actually present. In 2006 it was reported that upon review of "over 500 radical prostatectomy specimens from participants in the PCPT," no "significant" difference emerged in the Gleason scores of the two groups. See paper 54. Biopsies in the PCPT yielded about 2,000 findings of cancer. I assume most of those cases led to radical prostatectomies. As I have been informed, similar Gleason scores among radical prostatectomy specimens don't necessarily prove anything, as choices of treatment were not standardized from one medical center to another, and more aggressive cancers tend to be treated with radiation, not surgery.

41 "This, in fact, was observed": see paper 38.

41 "cancers that were never likely to produce morbidity or mortality": see paper 14.

42 "the balance of probability was that inoculation saved lives": Roy Porter, *The Enlightenment* (London: Macmillan, 1990), p. 66.

42 to compare the mortality statistics of the vaccinated German army against those of the other armies of Europe. Osler, *Evolution of Modern Medicine*, p. 129.

42 "no simple ratio that would allow one to weigh an increased life expectancy against a risk of immediate death": Wootton, *Bad Medicine*, pp. 154–155.

43 "the toxicity of many new agents has been discovered only after they were in regular use": Miles Weatherall in *Cambridge Illustrated History*, p. 275.

43 it was beginning to be studied for prostate cancer prevention. See paper 9.

43 it was the first synthetic therapy for cancer at all. See paper 33.

45 "the proportion of diagnosed tumors of no risk to the patient may be as high as seven out of eight": see paper 10.

45 "little or no residual cancer at radical prostatectomy after an initial diagnosis of minute cancer on needle biopsy": see paper 73.

45 "increased detection due to reduced gland volume contributed to the finasteride-associated increase in high-grade disease": see paper 44.

46 "may have been due to other causes besides truly induced aggressive disease": see paper 44.

46 just too dangerous to use preventively. Another piece of suggestive circumstantial evidence in finasteride's favor is that those in the finasteride group who underwent an investigative biopsy at the end of the PCPT showed virtually the same absolute number of high-grade tumors as the placebo group.

46 "the evidence does not rule out that finasteride may have induced high-grade prostate cancer in some men": see paper 44.

46 RAPID: Rapid Access to Preventive Intervention Development.

47 "are always worried about missing something and having someone else find it": Welch, *Should I Be Tested for Cancer?*, p. 118.

47 where the incidence of prostate cancer runs some *seven times* higher than in England and Wales. See paper 74.

48 when the question of using a powerful but questionable and poorly understood chemopreventive drug stands before them. Not all the finasteride authors are American, though.

48 "often follow the safest approach and recommend aggressive therapy": Peter Albertsen in Ian Thompson, Martin Resnick, and Eric Klein, eds., *Prostate Cancer Screening* (Totowa, N.J.: Humana Press, 2001), p. 28.

49 the rationale for the trial has been cogently questioned. See paper 49.

49 "first attracted interest for cancer prevention as an unexpected result" of clinical studies of their efficacy in preventing cardiovascular disease: see paper 20.

49 "provocative and unexpected" decreases in the incidence of certain cancers: see paper 20.

49–50 "these reductions increased with prolonged statin use": see paper 20.

50 inhibiting the less dangerous but not the more dangerous forms of the disease. "Finasteride may be more effective at preventing development of cancers with Gleason grade 2-6 than Gleason 7-10": see paper 14.

50 "Rarely is it emphasized that the definitive studies of prostate cancer screening have not been done": Otis Brawley in Thompson, Resnick, and Klein, *Prostate Cancer Screening*, p. 175.

51 hard to conceive that it would ever have been adopted as a treatment if it had been determined at the time to be a possible carcinogen. On the original clinical studies of Proscar, see paper 53. In these trials the effects of finasteride were found to be reversible, which implies that men would have to take the drug indefinitely to continue to enjoy its benefits. But of course the long-term consequences of taking a drug that had been in existence only for a few years could not be known.

51 the issue "clearly requires further study": *Wall Street Journal*, June 24, 2003.

51 Certain compounds of interest in the chemoprevention of prostate cancer have both estrogen and antiestrogen effects. See paper 7.

51 "Some agents can be preventive and carcinogenic in the same organ": see paper 41.

3. Sister Drugs: Finasteride and Tamoxifen

53–54 "account for about the same number of deaths per year in this country": Stewart Justman, *Seeds of Mortality: The Public and Private Worlds of Cancer* (Chicago: Ivan R. Dee, 2003), p. 47.

54 "but we can't say just which ones": Welch, *Should I Be Tested?*, p. 65.

54 "This suggests that prostate cancer prevention is also possible": see paper 68.

54 Certain prostate cancer patients have themselves been treated experimentally with tamoxifen. See paper 21.

55 "estrogen receptor-negative disease in the case of tamoxifen": see paper 66.

55 "no simple take-home message": Lawrence Altman, "Researchers Find the First Drug Known to Prevent Breast Cancer," *New York Times*, April 7, 1998.

55 "a widespread resistance to their use for preventing breast or prostate cancer": see paper 42. See also Scott Lippman, "The Dilemma and Promise of Cancer Chemoprevention," *Nature Clinical Practice: Oncology*, October 2006: 523.

56 the latter risk is considered "significant" because the women are healthy. DeGregorio and Wiebe, *Tamoxifen and Breast Cancer*, p. 50.

56 "treatment with tamoxifen stimulates the growth of a class of aggressive breast cancer tumors": Richard Stone, "NIH Fends Off Critics of Tamoxifen Study," *Science*, October 30, 1992: 734.

56 "Beware This Breakthrough!": *Time*, April 20, 1998.

56 "of 23 reported uterine cancers associated with tamoxifen, only ten had a good prognosis": DeGregorio and Wiebe, *Tamoxifen and Breast Cancer*, p. 51.

56 "a relatively unfavorable prognosis related to histology . . . and higher stage": see paper 16.

57 "compared with three months for the PCPT": see paper 19.

57 neither protested nor reported with any degree of follow-up. Some definite reluctance to proceed with a full-scale, Phase III clinical trial of finasteride appears in paper 18, a "panel consensus statement" published in 1992.

58 "unpleasant or even potentially life-threatening side effects": DeGregorio and Wiebe, *Tamoxifen and Breast Cancer*, p. 74.

58–59 "Women whose breast cancer risk is sufficiently high . . . would be candidates for the drug": see paper 23.

59 "the potential risks and side effects, which now seem virtually inescapable, of taking an effective chemopreventive agent": see paper 42.

59 "they are most often curable by hysterectomy and the mortality rate is minimal": see paper 23.

60 "using the number of adverse events to cancel out an equal number of tumors prevented is questionable": see paper 70.

60 "sufficient information to determine which women have a risk high enough to outweigh tamoxifen's potential hazards": DeGregorio and Wiebe, *Tamoxifen and Breast Cancer*, p. 85.

61 "one group received streptomycin whereas a control group was treated with traditional methods": Porter in *Cambridge Illustrated History*, p. 201; cf. p. 275.

61 "This allows those women on placebo to consider taking tamoxifen": see paper 23.

61 "extremely unlikely to change with additional diagnoses of prostate cancer and end-of-study biopsy results": see paper 67.

62 "may be more problematic in younger women": see paper 16.

4. Specific Harms and General Benefits

63 "men diagnosed with low-grade prostate cancer are more likely to be alive at 10 years compared with men without prostate cancer": see paper 69.

63 which makes the disease seem like a major public-health benefit. See Richard Richards's response to paper 74.

63 all warning signs, tests, and numerical measures unreliable in one degree or another. "At present there are no biologic, clinical, pathologic, or radiographic markers that allow the prediction of biologic significance of an individual tumor": see paper 36

63 "without ever having experienced a disease-related symptom": see paper 36.

64 the available information on the role of steroid hormones in the genesis of prostate cancer is "often contradictory": see paper 8.

64 Finasteride's most vigorous partisans hesitate to claim that it will actually save lives. As for tamoxifen, see paper 16.

64 "biologically active substances are likely to have adverse, perhaps unforeseen effects": see paper 42.

64 "commitment to the health of the nation and, indeed, of humanity supplied the implicit justification for many medical harms to individuals": Virginia Sharpe, "Why 'Do No Harm'?" *Theoretical Medicine* 18 (1997): 198.

65 "very aggressive and fast-growing, and it can be deadly": David G. Bostwick, et al., *Prostate Cancer: What Every Man—and His Family—Needs to Know* (New York: Villard, 1999), p. 66. Dr. Bostwick is a prominent figure in prostate cancer medicine.

66 "Speak of what is to happen tomorrow, and he will lend you attention": David Hume, *A Treatise of Human Nature* (Harmondsworth, Middlesex: Penguin, 1985), p. 475.

66 "People are generally motivated only by the prospect of a benefit . . . an individual's health next year is likely to be much the same": Rose, *Strategy of Preventive Medicine*, pp. 13, 105. Writing in 1992, Rose opposed the use of cholesterol-lowering drugs on the grounds that their long-term effects were unstudied and unknown. "The over-use of drugs is a constant danger in preventive medicine and a near-inevitable consequence of mass screening" (p. 112).

66 "the common mode of thought, a mode half good, half bad, but which there is no hope that men will reform in themselves": D'Alembert in Bradley, ed., *Smallpox Inoculation: An Eighteenth Century Mathematical Controversy*, p. 61.

67 "The mental anguish and concern caused by cancer screening should not be underestimated": see paper 10.

68–69 "distrusting ourselves as not being equal to deciding": *Complete Works of Aristotle*, ed. Jonathan Barnes (Princeton: Princeton University Press, 1984), II, 1756. The *Nicomachean Ethics* is translated by W. D. Ross, revised by J. O. Urmson.

69 "an individualized assessment of prostate cancer risk and risk of high-grade disease for men who undergo a prostate biopsy": see paper 63.

70 "ignores the real impact of a prostate cancer diagnosis in the United States today": see paper 30; see also papers 69 and 9.

70 15 cancers prevented per thousand men per seven years. See paper 19.

71 "the opportunity to never have to face the diagnosis in the first place (i.e., prevention)": see paper 14.

72 14.7 percent and 10.8 percent on the finasteride and placebo sides. See paper 3.

73 some think it helped skew the data of the PCPT against finasteride. See paper 3.

73 "has a higher-than-average risk of the disease, has urinary symptoms that may be relieved by finasteride, and is not sexually active": see paper 76.

73 An economic analysis of the finasteride question presumes a fifty-year-old man who takes the pill for twenty years. See paper 60.

73 "approximately 30 percent of men between the ages of 30 and 50 years have histologic cancers in their prostate": see paper 7.

73 "Complicating matters, finasteride also caused loss of libido and impotence": *Time*, July 7, 2003.

74 "to ensure that participants remained involved in the trial and were committed to getting their end-of-study biopsy": see paper 26.

74 close to 25 percent refused. See papers 27 and 45.

75 "Then we will witness the great clash of molecules": James Goodwin, *The Lancet*, April 28, 2001: 1376.

76 Reports tell of men breaking their five milligram pills of Proscar into fifths to get around the inflated price of one milligram pills of Propecia. *New York Times*, March 19, 1999. Then again, the University of California at Berkeley Wellness Letter (March 2003) has recommended splitting a one milligram pill of Propecia into fifths.

76 even though finasteride was present only in minute amounts. Lawrence K. Altman in the *New York Times*, June 23, 1992. Significantly, the article begins, "PROSCAR is an important new drug to treat symptoms of enlarged prostate glands, but as with any new drug the full range of its adverse effects remains to be assessed."

76 "Trials of orchiectomy were done in the 1940s on men with metastatic prostate cancer with evidence of . . . improvement": see paper 3.

5. Calculation of Harm

77 "become more and more insignificant in proportion as they occur less frequently": cited in E. Ashworth Underwood, "The History of the Quantitative Approach in Medicine," *British Medical Bulletin* 7 (1951).

78 "But the physician relieves them one by one: the legislator by millions at a time": cited in David Spadafora, *The Idea of Progress in Eighteenth-Century Britain* (New Haven: Yale University Press, 1990), p. 178.

78 "'cure' the defective mechanism of the human frame and the human mind": Janet Semple, *Bentham's Prison: A Study of the Panopticon Penitentiary* (Oxford: Clarendon Press, 1993), p. 153.

78 "The heavier side of the scale determined the action": Richard Altick, *Victorian People and Ideas* (New York: Norton, 1973), p. 118.

78 "remnant of the old nobility of medicine": Jonsen, *New Medicine*, p. 75.

79 "each is inspired and awed by the wisdom and nobleness of his or her chosen profession": see paper 56.

79 arguments that he thought served only to keep an ailing body politic ailing. *Bentham's Handbook of Political Fallacies*, ed. Harold A. Larrabee (New York: Apollo, 1971), p. 43.

80 "Unless this could be proved, he would account the infliction of punishment unwarrantable": *Mill's Essays on Literature and Society*, ed. J. B. Schneewind (New York: Macmillan, 1965), p. 248.

80 "begins all his inquiries by supposing nothing to be known on the subject": *Mill's Essays on Literature and Society*, p. 255.

81 of these 1,000 men 60 "would have been spared the diagnosis of cancer": see paper 30.

82 "pointless nostalgia": see paper 43.

82 "for the good of society, and not necessarily for their personal benefit": Pamela Clark and Paul Leaverton, "Scientific and Ethical Issues in the Use of Placebo Controls in Clinical Trials," *Annual Review of Public Health* 15 (1994): 31.

82 "until my asking and his answering become a pure waste of breath": G. J. Warnock, *The Object of Morality* (London: Methuen, 1971), p. 33.

83 "The loss of a happy life for the first infant is outweighed by the gain of a happier life for the second": Peter Singer cited in Jonsen, *New Medicine*, p. 108.

83 life-threatening or fatal complications in 5 percent of all hospitalized patients. Sharpe and Faden, *Medical Harm*, p. 63.

83 five deaths for every ninety-five surviving patients do not make for good medicine. Cf. Hannah Arendt, *Crises of the Republic* (New York: Harcourt Brace Jovanovich, 1972), p. 37: "If . . . it can be calculated that the outcome of a certain action is 'less likely to be a general war than more likely,' it does not follow that we can choose it even if the proportion were eighty to twenty, because of the enormity and *incalculable quality* of the risk" (emphasis in the original).

84 "they are not *uniquely* beneficent toward me": Warnock, *Object of Morality*, p. 33.

84 the surveillance mechanisms and wholesome regimen of his ideal prison, the Panopticon. See Stewart Justman, *The Psychological Mystique* (Evanston: Northwestern University Press, 1998).

85 A founder of the Statistics Society: Jonsen, *New Medicine*, p. 113.

85 "which therefore reduced the overall happiness of the community": John Dinwiddy, *Bentham* (Oxford: Oxford University Press, 1989), p. 25.

86 "incite the commission of a greater crime to prevent the detection of a less": *Rambler* No. 114, April 20, 1751.

86 "versus 27 low/intermediate grade prostate cancers plus 18 high-grade prostate cancers, and 15 cases prevented": see paper 70.

87 "more than 300,000 person years would be saved . . . assuming no change in the rate of high-grade prostate cancers": see paper 70.

87 "The potential detrimental effects . . . would be outweighed by a reduction in incidence": see paper 69.

88 "supplied the implicit justification for many medical harms to individuals": Sharpe and Faden, *Medical Harm*, p. 82.

88 "159,680 person-years still would have been saved, representing a positive benefit to society": see paper 69.

89 "whether the utilitarian rationale for medical harm was still valid": Sharpe and Faden, *Medical Harm*, p. 66.

90 "This is certainly the case when the target is to eradicate a disease requiring the vaccination of millions of people": see paper 51.

90 "to fancy himself a dead Iron-Balance for weighing Pains and Pleasures on was reserved for this his latter era": *A Carlyle Reader*, ed. G. B. Tennyson (New York: Modern Library, 1969), p. 276.

91 "the direct begetter . . . of the science of preventive medicine": Benjamin Spector, "Jeremy Bentham, 1748–1832, His Influence Upon Medical Thought and Legislation," *Bulletin of the History of Medicine* 37 (1963): 30. On the connection between enlightened penal practices and prevention, see Rose, *Strategy of Preventive Medicine*, p. 30: "Just as in an enlightened penal system the treatment of offenders is seen as an opportunity to make future offences less likely, so also in medicine the clinician who treats a sick person has the opportunity to use the occasion for prevention as well as cure."

91 "very safe and very quiet": Semple, *Bentham's Prison*, pp. 123, 112.

91 places Bentham's design for the Panopticon in the utopian tradition. I refer to Semple's *Bentham's Prison*. Bentham dreamed of an amusement park worthy of More's Utopia on the grounds of the Panopticon, including a tavern serving various gorgeous varieties of water.

91 Bentham liked the word "liberty" less than "security": Semple, *Bentham's Prison*, p. 22.

91 If the health of the people is the supreme law . . . as Bentham among many others seems to have believed: Spector, "Jeremy Bentham, 1748–1832": 42.

92 it is the "mission" of medicine to regulate human beings in their own interest. See paper 17.

92 "proper isolation of the sick": see paper 17. On the control of typhus in New York at the turn of the twentieth century, see Howard Markel, "When Germs Travel," *American Scholar*, Spring 1999.

92 "Maybe freedom suffers less if it is attacked from both sides": Rose, *Strategy of Preventive Medicine*, p. 118.

93 "at the population level": see paper 70.

93 unless they were steered into it? "How risk information is presented, worded, and framed may affect its interpretation": see paper 16.

93 "a decision by the physician to give the drug for cancer prevention and then to watch the patient carefully": see paper 76.

93 "After 7 years of [the PCPT], all subjects with a history of normal screening examinations were given a biopsy of the prostate": see paper 11.

93 "placing all black men on 5ARIs may be premature given the current level of evidence": see paper 24.

93 only so many physicians will recommend it, and only so many patients will accept it. See paper 61.

94 "the conservatism of the doctrine of the desirability of pain": Youngson, *Scientific Revolution in Victorian Medicine*, p. 95; see also p. 112.

94 will end up writing prescriptions for the drug willy-nilly because patients will demand it. Similarly, while ads with images of middle-aged gentlemen at play prompt men to talk to their doctor about BPH medication like dutasteride, there are no ads urging men to consult with their doctor to see if a drug to prevent prostate cancer is right for them.

94 "to behave as if their ends are less ultimate and sacred than my own": Isaiah Berlin, *Four Essays on Liberty* (London: Oxford University Press, 1970), p. 137. According to a 1976 paper on preventive medicine published by the British government, "Preventive medicine has always involved some limitation of the liberty of the individual, often for his own good, often for that of the community, and often for both." Cited in Howard Leichter, *Free to Be Foolish: Politics and Health Promotion in the Untied States and Great Britain* (Princeton: Princeton University Press, 1991), p. 82.

96 "many studies attempted the quantification of results that would allow physicians to do just that": Jonsen, *History of Medical Ethics*, p. 83.

97 "These practitioners were willing to incur danger in order to prevent suffering, but only up to a moderate limit": Martin Pernick, "The Calculus of Suffering in Nineteenth-Century Surgery," *The Hastings Center Report*, April 1983: 30.

97 In Britain only two or three deaths in a year as a result of chloroform "would have made a strong impression": Youngson, *The Scientific Revolution in Victorian Medicine*, p. 80.

97 "one death in ten thousand cases is sufficient to condemn chloroform on moral grounds": Pernick, "The Calculus of Suffering in Nineteenth-Century Surgery": 36.

97 "anaesthetized approximately 15,000 patients between them, and did not have a single fatality": Youngson, *The Scientific Revolution in Victorian Medicine*, p. 214.

6. Medical Knowledge and Medical Ignorance

99 "make prevention of the disease an appealing strategy": see paper 67.

99 *"the significance of these cancers is uncertain"*: see paper 30. My emphasis.

99 *"may* also significantly inhibit the development and progression of prostate cancer": see paper 2. My emphasis.

99 *"it is possible* that greater suppression of DHT could lead to better efficacy in preventing prostate cancer": see paper 2. My emphasis.

99 *"We do not know* the degree to which increased risk of detection of cancers resulted in underestimating the risk reduction for finasteride": see paper 36. My emphasis.

99 *"It is uncertain whether it can ever be determined"*: see paper 14. My emphasis.

99 *"the profound complexity of carcinogenesis"*: see paper 41. My emphasis.

100 "the precise mechanisms by which androgens affect this process and the possible involvement of estrogenic hormones are not clear": see paper 8.

100 "understanding of the pathophysiology of BPH remains incomplete": see paper 71.

101 "accumulated knowledge, norms, and ideals handed down by previous generations": Edward Shils, *Tradition* (Chicago: University of Chicago Press, 1981), pp. 10–11.

101 a hyperbolic version of this already painfully simplistic story. See Stewart Justman, *Fool's Paradise: The Unreal World of Pop Psychology* (Chicago: Ivan R. Dee, 2005).

102 "at least an understanding of disease mechanisms and drug action *kept them from doing harm"*: Shorter in *Cambridge Illustrated History*, p. 128.

102 "Socrates counseled epistemological modesty": Colin McGinn, *Shakespeare's Philosophy: Discovering the Meaning Behind the Plays* (New York: HarperCollins, 2006), pp. 4–5. The last words of Socrates concern his debt to Asclepius, the god of healing. No one knows what they mean.

102 "the alarming fact that the same world can appear differently to different observers": Karl Mannheim, *Ideology and Utopia*, tr. Louis Wirth and Edward Shils (New York: Harcourt, Brace & World, 1936), p. 6.

103 "the wisest man there ever was . . . medicine is received like geometry": Michel de Montaigne, *Apology for Raymond Sebond*, tr. Roger Ariew and Marjorie Grene (Indianapolis: Hackett, 2003), pp. 62, 103, 118, 121.

103 Our very bodies, our physical selves, refute the human pretension to knowledge. Cf. Socrates' rather modest view of medicine, in Book 3 of the *Republic*, as an art capable of doing some repair work on the human body but not of restoring to health those not fundamentally healthy to begin with.

103 "threatening me with suffering and then with imminent death": Michel de Montaigne, *The Essays: A Selection*, tr. M. A. Screech (London: Penguin, 1993), p. 394.

103 "keeps her processes absolutely unknown": Montaigne, *Essays*, p. 400.

104 "we may derive rules of medicine more certain than those which we have had up to the present": René Descartes, *Discourse on Method*, tr. Laurence Lafleur (Upper Saddle River, N.J.: Prentice Hall, 1956), p. 78.

104 "Our business here is not to know all things, but those which concern our conduct": John Locke, Introduction to *An Essay Concerning Human Understanding* (New York: New American Library, 1974), p. 65.

105 to mark that field in which we can profitably work. So influential was Locke's argument concerning the limits of human understanding that we seem to hear it in the more febrile voice of Rousseau decades later. Convinced that Nature has drawn a "heavy veil" over her activities, so that we do not and cannot know things in themselves (a position Kant would take as well), Rousseau maintains that "the reasonable and modest man" is one "whose practiced but finite understanding is aware of and accepts its limitations." Cited in Jean Starobinski, *Jean-Jacques Rousseau: Transparency and Obstruction*, tr. Arthur Goldhammer (Chicago: University of Chicago Press, 1988), p. 76. The modest man is modest because he recognizes his own ignorance, and his understanding "practiced" because he also recognizes, with Locke, that ignorance does not excuse idleness.

105 "may be carried much further than it hitherto has been": Locke, *Essay*, p. 333.

105 "to discern how far our knowledge does reach": Locke, *Essay*, p. 333.

105 "His treatment was always simple and safe": Kenneth Dewhurst, *John Locke (1632–1704), Physician and Philosopher: A Medical Biography* (London: Wellcome Historical Medical Library, 1963), pp. 296, 297, 311.

106 "So that there is nothing left for a Physician to do, but observe well": cited in Dewhurst, *John Locke*, p. 310.

106 "whether rhubarb will purge or quinquina cure an ague can be known only by experience": cited by William Osler, "John Locke," in *Collected Essays*, Vol. III (New York: Gryphon, 1996), p. 197.

106 "arguably the first effective specific drug": Roy Porter, *The Greatest Benefit to Mankind: A Medical History of Humanity* (New York: Norton, 1997), p. 230.

106 "it was necessary to remove a great deal of rubbish": cited by Weatherall in *Cambridge Illustrated History*, p. 274.

106 Lind was perhaps the first to advocate the use of a placebo for control purposes in a clinical trial. See Clark and Leaverton, "Scientific and Ethical Issues in the Use of Placebo Controls."

107 "and a subject so sacred as a human life": cited in Sharpe and Faden, *Medical Harm*, p. 9.

107 "a sick man who did not consult a physician had a better chance of surviving than one who did": Peter Gay, *The Enlightenment: An Interpretation*; Vol. II: *The Science of Freedom* (New York: Alfred A. Knopf, 1969), p. 19.

107 "the remote causes of disease were . . . beyond the range of human understanding": Dewhurst, *John Locke*, p. 35.

107 the displacement of doctrinaire knowledge by the spirit of inquiry. Gay, *The Enlightenment*, p. 22.

108 "that which teaches us where knowledge leaves off and ignorance begins": Holmes, *Medical Essays*, p. 211.

108 "the study of which should always be approached with humility and reverence": cited in Youngson, *Scientific Revolution in Victorian Medicine*, p. 140.

108 "the ignorance that exists as a precondition for scientific progress": see paper 29.

109 "honest admission of ignorance": see paper 58.

109 "The numbers are small": CNN, June 24, 2003.

112 "and (6) that PSA testing is controversial": Mark Litwin and Kristen Reid, in *Prostate Cancer Screening*, eds. Thompson, Resnick, and Klein, p. 193.

113 "This, in fact, was observed": see paper 38.

114 "This was indeed the case in the PCPT": see paper 3.

114 "would be expected to contribute to greater detection of all grades of prostate cancer with finasteride": see paper 65.

114 "the Aug. 16 issue of the *Journal of the National Cancer Institute*": Med-Page Today, August 16, 2006.

115 The final figure turned out to be four times that. The disparity is well analyzed in paper 47.

115 "unsurpassed by any other tumor in the history of modern public health statistics": Thompson, Resnick, and Klein, *Prostate Cancer Screening*, p. v. Significantly, even the unprecedented skyrocketing of detected cancers with PSA testing is represented by Dr. Thompson and a co-author as something retrospectively predictable. "With the first proliferation of PSA testing, it would be expected that a large group of men with previously undiagnosed prostate cancer would be identified" (p. 162). Earlier in the same paper, however, the authors bring out the actually astonishing character of the explosion of cancer diagnoses. With the advent of PSA testing, "the result was an extraordinary surge in the number of prostate tumors detected. Indeed, never in the experience of the monitoring system set up by the National Cancer Institute . . . had such an incidence in disease incidence been witnessed" (p. 158). The same set of events is described now as unprecedented and extraordinary, now as predictable. A 1987 editorial in the *New England Journal of Medicine* announcing the arrival of PSA envisions its use to monitor recurrence in patients who have undergone radical prostatectomy. No inkling or intimation of the epidemic of diagnoses that followed the inauguration of

PSA testing appears. See R. F. Gittes, "Prostate-specific Antigen," *New England Journal of Medicine*, October 8, 1987.

115 "the revolution in prostate treatments that followed the discovery of pseudohermaphrodites": see paper 46.

116 "but would not affect libido, potency, or male musculature": see paper 53.

116 "5 alpha-reductase inhibitors and retinoids may enhance rather than slow down the progression of prostatic cancer": see paper 6; see also paper 47.

116 "contrary to the expectations of many experts": see paper 27.

116 "no serious drug-related safety concerns have arisen": see paper 68.

117 "critics claimed that this figure was too high": DeGregorio and Wiebe, p. viii.

117 "did not exceed expectations": Altman, *New York Times*, April 7, 1998.

117 "Such events had not been prospectively defined": see paper 35.

118 confounding the illusion that events unfold in ways that could only have been expected. See Stewart Justman, *Literature and Human Equality* (Evanston: Northwestern University Press, 2006).

118 "purported Truth becomes fallible opinion; there are no final truths": Gary Saul Morson, *Narrative and Freedom: The Shadows of Time* (New Haven: Yale University Press, 1994), p. 268.

119 "the effect of a given therapeutic intervention on a given patient is always to some extent uncertain": see paper 29.

7. Prevention Gone Too Far

120 "So long as private property remains, there is no hope at all of effecting a cure and restoring [European] society to good health": Thomas More, *Utopia*, tr. Robert M. Adams (Cambridge, UK: Cambridge University Press, 1993), p. 39.

120 "that no one shall value gold and silver": *Utopia*, p. 62.

121 "no danger from internal strife, which alone has been the ruin": *Utopia*, p. 110.

121 "could only tell women what to avoid, not what to buy": Amy Allina and Cindy Pearson, "Pills, Prevention, and Profits," *Multinational Monitor*, September 1999: 2–3.

122 "how dearly nature makes us pay for the contempt we have shown for her lessons": Note I to *Discourse on Inequality*, tr. Maurice Cranston (Harmondsworth, Middlesex: Penguin, 1984), p. 149.

122 no one suggests that men simply skip screening and get themselves biopsied annually. . . . See paper 10.

122 "as many as 32%": see paper 24.

122 "shifting increasingly toward the concerns of society": see paper 43.

123 "academics, physicians, policymakers, and the general public": Leichter, *Free to Be Foolish*, p. 76.

123 "legislative and public health policies; [and] health insurance policies": see paper 41.

123 "universal" monitor: Porter, *Greatest Benefit*, pp. 632–633.

124 "depression, restless leg syndrome, and sexual dysfunction": Welch, et al., "What's Making Us Sick," *New York Times*, January 2, 2007.

125–126 "She cries not with hopeless tears, but with tears of joy and relief and release": *Divine and Human and Other Stories by Leo Tolstoy*, tr. Peter Sekirin (Grand Rapids: Zondervan, 2000), pp. 43–44.

126 It is one thing for Tolstoy's own Ivan Ilych to remember his mother on his deathbed and mourn the loss of his childhood. See my discussion in *Seeds of Mortality*.

127 patients who choose this course "fare as well as those who do not in the short and long term": see paper 62.

8. Breakthrough: "The Present Is Obsolete"

128 "and to the contamination of the whole field by big money": Porter, *Greatest Benefit*, p. 579.

129 "The present is obsolete": Charles Coltman, *The Oncologist* 2 (1997): 206.

129 "substitute an era of cold fact for the present era of heated opinion": Peter Albertsen in *Prostate Cancer Screening*, eds. Thompson, Resnick, and Klein, p. 43. The researcher is Willet Whitmore.

130 "millions of women will probably be advised to take it for the rest of their lives": "Powerful Hormone to Be Tested in War to Prevent Breast Cancer," *New York Times*, September 14, 1991.

130 PSA . . . was originally of interest as a tool in rape investigations: see paper 14.

130 It was approved by the FDA in 1986 . . . as a way of monitoring the progress of prostate cancer: see paper 22.

130 "miracle cure for rheumatoid arthritis": Hardy, *Health and Medicine in Britain*, p. 156.

131 "opened up a whole new dimension of popular, patient-driven illness": Hardy, *Health and Medicine in Britain*, p. 158.

131 "these undeniable producers of dramatic effects are the exception rather than the rule, even in these halcyon days of antibiotics": see paper 31.

132 "many men who would have qualified as controls . . . are now known to have prostate cancer as a result of PSA screening": see paper 50.

132 "discordant interpretations were arbitrated by a referee pathologist": see paper 26.

132 "The molecular hallmarks of cancer development . . . are frequently present in both cancer and precancer": see paper 1.

132 "across a broad spectrum of neoplastic stages ranging from premalignant to invasive disease": see paper 66.

133 "therapy and prevention are also blurring at the clinical level": see paper 41.

134 "anti-cancer drugs generally have proved more effective in combinations than singly": Lawrence Altman, "The Doctor's World," *New York Times*, May 26, 1998.

134 "much as insulin keeps diabetics alive without curing diabetes": Altman, *New York Times*, May 26, 1998.

134 "There's no sure way to prevent it until now": CNN, June 24, 2003.

136 much of rhetoric concerns probabilities, or things that happen "for the most part": Aristotle, p. 2157.

136 "the long tradition shapes medicine": Jonsen, *New Medicine*, p. 8.

9. "I Will Abstain from Harming Any Man"

138 "only if they could be rendered completely non-poisonous": Pernick, "The Calculus of Suffering in Nineteenth-Century Surgery": 34.

138 "the paradox of the murderous healer was the subject matter of a great many rhetorical exercises": Sharpe and Faden, *Medical Harm*, p. 6.

138 "the presumption in favor of poisoning . . . is not extinct in those who are entrusted with the lives of their fellow-citizens": Holmes, *Medical Essays*, p. 256.

140 "she had to kill some rats that were keeping her from falling asleep": Gustave Flaubert, *Madame Bovary*, tr. Mildred Marmur (New York: New American Library, 1964), p. 292.

140 "lagged well behind medicine in gaining academic recognition": Dewhurst, *John Locke*, p. 307.

140 "severance between medical and surgical knowledge": George Eliot, *Middlemarch* (Boston: Houghton Mifflin, 1968), p. 108.

141 "while [their] assistant pointed to the organs alluded to and a dissector did the knifework": Porter in *Cambridge Illustrated History of Medicine*, p. 154.

141 "alchemy, astrology, grammar, lexicography, logic, scholastic disputation, arithmetic, algebra, and history": *The Arabian Nights*, tr. Husain Haddawy (New York: Alfred A. Knopf, 1990), p. 255.

142 "frequently there is more danger from the physician than from the distemper": Shorter in *Cambridge Illustrated History*, p. 124.

142 "Morphine was extracted from opium, quinine from cinchona, but people went on dying, more or less as before": Wootton, *Bad Medicine*, p. 183.

143 "it is important to remain focussed on the effectiveness, and the toxicology, of any medication that appears on the market": see paper 72.

144 "mankind would not longer be enslaved by tradition": Shils, *Tradition*, p. 21.

145 contemporary surgical fads such as "correcting" stutters by cutting the root of the tongue. See Porter, *Greatest Benefit*, pp. 362–363, 383.

147 information picked up from a book borrowed for the occasion. In this sense his imagination, like his wife's, is seduced by reading.

148 "Locke has unfolded to man the nature of human reason as a fine anatomist explains the powers of the body": Voltaire, *Philosophical Letters*, tr. Ernest Dilworth (Indianapolis: Library of Liberal Arts, 1961), pp. 53–54.

149 "we know that it would have been out of character for him to do so": Frederick Crews, *Follies of the Wise: Dissenting Essays* (Emeryville, Calif.: Shoemaker and Hoard, 2006), pp. 18–19.

149 first used in 1924 in a textbook of psychiatry: Sharpe and Faden, *Medical Harm*, p. 61.

150 "the desperate gullibility of patients prepared to believe that each and every organ could spawn dozens of defects": Porter in *Cambridge Illustrated History*, p. 113.

151 as one French doctor did in a treatise on hysteria: Stephen Heath, *Gustave Flaubert; Madame Bovary* (Cambridge, UK: Cambridge University Press, 1992), p. 97.

151–152 "They said it was a study that would do you good": see paper 4.

152 "more potential harm for the patient than potential benefit": James Jones, *Bad Blood: The Tuskegee Syphilis Experiment* (New York: Free Press, 1993), p. 7.

152 many diseases were not to be treated: Shorter in *Cambridge Illustrated History*, pp. 138, 141.

152 "some studies . . . suggested that syphilis did not always need to be treated—that it could often remain quiescent, especially in blacks": see paper 4.

153 "the longest observational study in medical history": see paper 4.

153 "substantial distrust of government-supported research": see paper 48.

153 "little harm was done by leaving the men untreated": Jones, *Bad Blood*, p. 7.

154 "Not treating them had become routine": Jones, *Bad Blood*, p. 164.

154 reminding physicians of their duty "to prevent harm and to heal the sick whenever possible," and citing the Hippocratic Oath: Jones, *Bad Blood*, p. 11.

155 "telling his story to someone he trusts as a 'healer'": Edward Shorter, *Doctors and Their Patients: A Social History* (New Brunswick, N.J.: Transaction, 1991), p. 252.

155 "an enormous role in the practice of medicine": Shorter, *Doctors and Their Patients*, p. 151.

156 "*the absence of this one means may prove fatal*": Hooker, *Physician and Patient*, p. 397.

156 "its indiscriminate employment by ignorant persons should be prevented by law": Osler, "Medicine in the Nineteenth Century" in *Collected Essays*, Vol. III, pp. 273, 275.

156 "That was as much help to them as a dose of medicine": Jones, *Bad Blood*, pp. 164–165.

10. "Do *No* Harm?"

158 "chemopreventive agents must be entirely free of any side effects": see paper 18.

158 To administer a risky drug to a large population over a long time is asking for trouble. See paper 12.

158 "*without injuring the soft tissues of the patient*": Sherwin Nuland, *Doctors: The Biography of Medicine* (New York: Vintage, 1995), p. 364; my emphasis.

158–159 "Childhood vaccines against measles, rubella, and pertussis, however, increase life expectancy by about .1 month each": See Yair Lotan, Jeffrey Caddedu, J. Jack Lee, Claus Roehrborn, and Scott Lippman, "Implications of the Prostate Cancer Prevention Trial: A Decision Analysis Model of Survival Outcomes," *Journal of Clinical Oncology*, March 20, 2005.

159 a quarter of the strokes and a fifth of the heart attacks in the United Kingdom: Rose, *Strategy of Preventive Medicine*, p. 78.

159 "these women were never a real source of absolute danger in terms of the number of killers and the numbers killed": Vanessa McMahon, *Murder in Shakespeare's England* (London: Hambledon and London, 2004), p. 130.

160 "it could no doubt be rendered in Latin for those who love the quasi mystical authority of an ancient language": see paper 13.

161 "the best available response to the presenting symptoms": see paper 29.

161 "a significant health burden for men": National Cancer Institute news release, June 24, 2003.

161 "a greater incidence of more aggressive disease could outweigh any benefit from an overall reduction in cancer risk": see paper 45.

162 "would that we could only quantify these probabilities more precisely!": Lou Lasagna cited in paper 57.

162 "every individual will assign his own weights . . . and other indefinable values": see paper 36; see also paper 37.

163 they failed it. See paper 39.

164 "as the Institute of Medicine report asserts": see paper 56.

164 "most preventable [medical] harms are the result of system failures": Sharpe and Faden, *Medical Harm*, p. 6. The authors, however, would like to bureaucratize medical responsibility. They deem the practice of blaming single persons for medical errors self-defeating and obsolete, and concur that physicians suffer from perfectionism. It is ironic that a book that looks closely at the issue of medical harm can agree to this extent with physicians who make very short and breezy work of an epidemic of medical *deaths*.

165 *To Err Is Human: Building a Safer Health System*, eds. Linda T. Kohn, Janet M. Corrigan, and Molla S. Donaldson (Washington, D.C.: National Academy Press, 2000).

165 "The focus must shift from blaming individuals for past errors to a focus on preventing future errors by designing safety into the system": *To Err*, pp. 4–5.

165 "'First do no harm' is an often quoted term from Hippocrates": *To Err*, p. 3.

166 "to accept error as normal" and embrace it "as an opportunity to learn": *To Err*, p. 179.

166 Compared to this bit of doubletalk, the principle Do No Harm possesses an admirable clarity. Likening the medical industry to other industries, this report maintains that "the application of human factors [sic] in other industries has successfully reduced errors. Health care has to look at medical error not as a special case of medicine, but as a special case of error, and to apply the theory and approaches used in other fields to reduce errors and improve reliability" (p. 66). Medicine is not just one industry among others. It is distinguished by its unique concern with human illness and well-being, and its history and unique traditions, including the proscription of harm. A hospital is not an airline.

166 "It was frequently said that many deaths due to chloroform were hushed up or not reported" in Britain: Youngson, *Scientific Revolution in Victorian Medicine*, p. 80.

166 in 1974, a Senate investigation found that some twelve thousand Americans died each year as a consequence of unnecessary surgeries alone: Porter, *Greatest Benefit*, p. 687.

167 "Perhaps the ancient Hippocratic injunction, 'do no harm,' need not yield so easily to the demands of 'modern' medicine": Gillam and Bernstein, "Doing Harm": 82.

167 "there are actions one cannot perform; more precisely, would dread to perform": Philip Rieff, *The Feeling Intellect* (Chicago: University of Chicago Press, 1990), p. 323.

167 "is increasingly subverted, ignored, altered, reinterpreted" in practice: see paper 43.

167 someone who played like a lawyer with the cardinal precept of medicine. On the kinship of doctors and lawyers, see Plato, *Republic* Book 3: "What better proof can there be of a bad and disgraceful state of education than this, that not only artisans and the meaner sort of people need the skill of first-rate physicians and judges, but also those who would profess to have had a liberal education? Is it not disgraceful, and a great sign of want of good-breeding, that a man should have to go abroad for his law and physic because he has none of his own at home, and must therefore surrender himself into the hands of other men whom he makes lords and judges over him?" Tr. Jowett. The satiric imagination also sees doctor and lawyer as brethren.

168 "And here Bentham's conception of human nature stops": *Mill's Essays on Literature and Society*, ed. Schneewind, p. 262.

168 "There is a studied abstinence from any of the phrases, which, in the mouths of others, import the acknowledged existence of such a fact": ibid.

169 "The moral obligation to do no harm can be justifiably overridden but it can never be erased": Sharpe and Faden, *Medical Harm*, p. 85.

169 "and is proportionately less harmful than the condition for which the patient sought care": Sharpe and Faden, *Medical Harm*, p. 124.

169 "than to let the financial burdens of such care fall randomly where they may": see paper 29.

170 "It is the withdrawal of that support, as the nuclear family comes crashing down, that leaves us confused and frightened at the signals of our own bodies": Shorter, *Doctors and Their Patients*, p. 221.

171 "Indeed, in England that idea and the Victorian period began together": Walter Houghton, *The Victorian Frame of Mind* (New Haven: Yale University Press, 1957, p. 1.

171 "God forbid . . . that any member of the profession . . . should hazard it negligently, unadvisedly, or selfishly!": Holmes, *Medical Essays*, p. 167.

172 "It may seem a strange principle to enunciate as a first requirement in a hospital that it should do the sick no harm": cited in paper 57.

172 "an excruciating accusation": Nuland, *Doctors*, p. 247. The 1847 Code of Medical Ethics of the AMA does not cite the Hippocratic rule, though its preamble does affirm that "the duties of a physician were never more beautifully exemplified than in the conduct of Hippocrates, nor more eloquently described than in his writings."

172 many doctors in Britain were reluctant to report patients who died of puerperal fever. See Hardy, *Health and Medicine in Britain Since 1860*, p. 24.

172–173 "the revolution, which is now taking place in the practice of medicine": Hooker, *Physician and Patient*, pp. 219–220.

173 "he has been killing patients by allowing into their wounds microbes which he should have been destroying": Nuland, *Doctors*, p. 370.

173 the word "Victorian" has become an epithet. See Christopher Clausen, "The Great Queen Died," *American Scholar*, Winter 2001: 41–49.

173 "a rule of thumb that [feels] good": see paper 13.

174 "The Hippocratic ethic is dead": R. M. Veatch cited in Miles, *The Hippocratic Oath and the Ethics of Medicine*, pp. 125–126.

174 "the obligations of beneficence and nonmaleficence have been interpreted and exercised paternalistically": Sharpe and Faden, *Medical Harm*, p. 53.

175 "only about 15 per cent of medical interventions are supported by solid scientific evidence": see paper 58.

176 "One might expect . . . there might be less use of the drugs": see paper 15.

176–177 "provided that the Hippocratic injunction to do no harm is not violated, much good can be accomplished during the period between a new therapy's introduction and its scientific validation": Nuland, *Doctors*, p. 254.

177 "this probably explains his nihilistic approach to therapeutics": Thomas P. Duffy, "Sir William Osler Revisited," *Yale Journal of Biology and Medicine* 53 (1980): 222.

177–178 "a multiplicity of remedies the action of which is extremely doubtful": "Medicine in the Nineteenth Century" in *Collected Essays*, Vol. III, pp. 268–269.

179 "After three days of almost constant convulsions, Lovett died." Kenneth De Ville, *Medical Malpractice in Nineteenth-Century America: Origins and Legacy* (New York: New York University Press, 1990), p. 36.

179 "independently or in spite of the mercury": John Stuart Mill, *System of Logic* (New York: Harper and Brothers, 1846), p. 262.

11. The Finasteride Story: What Did Not Happen

181 "I am firmly convinced that this is the first step in conquering prostate cancer": University of Texas Health Science Center at San Antonio, news release, June 26, 2003.

182 nowhere to be found in the actual medical literature on finasteride. The collection of essays on prostate cancer screening co-edited by the same author did, however, strike a European reviewer as "evangelical": see paper 32.

182 "the magnitude of the increase in higher-grade tumors was still much less than the reduction in overall cancers": University of Texas Health Science Center at San Antonio, news release, June 26, 2003.

183 "finasteride reduces the risk of prostate cancer": see paper 74.

184 The doubts and dangers . . . are given their full weight in the pages of the *New England Journal of Medicine*: see paper 67.

185–186 "concerns regarding the higher prevalence of high-grade disease . . ." have blocked the drug's "widespread use": see paper 14.

186 "the men on Proscar who did develop cancer tended to get more deadly tumors, possibly because of the drug": *Business Week*, June 24, 2003 and June 7, 2004.

186 a director of the PCPT confessed that the study yielded "a mixed message": Charles Coltman as quoted in the *Wall Street Journal*, June 24, 2003.

186 "The increased risk of high-grade disease in the PCPT was due, *at least in part*, to improved detection, rather than the induction of high-grade disease": *MedPage Today*, August 16, 2006; see paper 65.

186 the study's designers did not adequately compensate for finasteride's effect on PSA: "Baldness Drug Can Mask Prostate Cancer Marker," Reuters, December 5, 2006. The researcher argued that "as men are on the drug longer, [the PSA multiplier of two] needed to be increased"; yet it was increased in the PCPT from 2.0 to 2.3. In addition, the greatest increase in high-grade tumors in the PCPT did not occur with prolonged use of finasteride, but early on, when its effect on the volume of the prostate was most pronounced.

187 "on balance, finasteride does not seem to be an attractive agent for the chemoprevention of prostate cancer": see paper 55.

187 "chemoprevention of prostate cancer is at hand": *Journal of the National Cancer Institute*, August 16, 2006: 1104.

187 "He claimed the cancers prevented by Finasteride were ones which would never have needed treatment anyway": "A New Method to Predict Prostate Cancer Recurrence Risk," news release, University of California at San Francisco, November 17, 2006. Dr. Scardino has informed me that in his view "Finasteride and other 5 alpha-reductase inhibitors have proven value in relieving the symptoms of BPH and in reducing the risks of urinary retention. Their value in reducing the risk of prostate cancer is less certain but very promising and I occasionally recommend [finasteride] to patients with very low risk cancers or a watchful waiting protocol."

188 "The Effects of 5 α-Reductase Inhibitors on the Natural History, Detection, and Grading of Prostate Cancer": see paper 3.

190 "The potential detrimental effects of an increased rate of patients who have prostate cancer with high-grade Gleason scores would be outweighed by a reduction in incidence": see paper 69.

Bibliography:
Medical Papers Cited in the Notes

1. Abbruzzese, James, and Scott Lippman, "The Convergence of Cancer Prevention and Therapy in Early-Phase Clinical Drug Development," *Cancel Cell* 6 (2004): 321–326.
2. Andriole, G., C. Roehrborn, C. Schulman, K. Slawin, M. Somerville, and R. Rittmaster, "Effect of Dutasteride on the Detection of Prostate Cancer in Men with Benign Prostatic Hyperplasia," *Urology* 64 (2004): 537–543.
3. Andriole, Gerald, David Bostwick, Francisco Civantos, Jonathan Epstein, M. Scott Lucia, John McConnell, and Claus Roehrborn, "The Effects of 5α-Reductase Inhibitors on the Natural History, Detection, and Grading of Prostate Cancer: Current State of Knowledge," *Journal of Urology* 174 (2005): 2098–2104.
4. Baker, Shamim, Otis Brawley, and Leonard Marks, "Effects of Untreated Syphilis in the Negro Male, 1932 to 1972: A Closure Comes to the Tuskegee Study, 2004," *Urology* 65 (2005): 1259–1262.
5. Berendes, Heinz, and Young Lee, "Suspended Judgment: The 1953 Clinical Trial of Diethylstilbestrol During Pregnancy: Could It Have Stopped DES Use?" *Controlled Clinical Trials* 14 (1993): 179–182.
6. Bosland, Maarten, "Possible Enhancement of Prostate Carcinogenesis by Some Chemopreventive Agents," *Journal of Cellular Biochemistry*, Supplement 16H (1992): 135–137.
7. Bosland, Maarten, "The Role of Estrogens in Prostate Carcinogenesis: A Rationale for Chemoprevention," *Reviews in Urology* 7 (2005): S4–S10.
8. Bosland, Maarten, "The Role of Steroid Hormones in Prostate Carcinogenesis," *Journal of the National Cancer Institute Monographs* 27 (2000): 39–66.

9. Brand, Timothy, Edith Canby-Hagino, A. Pratap Kumar, Rita Ghosh, Robin Leach, and Ian Thompson, "Chemoprevention of Prostate Cancer," *Hematology/Oncology Clinics of North America* 20 (2006): 831–843.

10. Brawley, Otis, and Barnett Kramer, "Cancer Screening in Theory and Practice," *Journal of Clinical Oncology* 23 (2005): 293–300.

11. Brawley, Otis, "Chemoprevention in Prostate Cancer," *Current Problems in Cancer* 28 (2004): 218–230.

12. Brawley, Otis, "The Potential for Prostate Cancer Chemoprevention," *Reviews in Urology* 4; Supplement 5 (2002): S11–S17.

13. Brewin, Thurstan, "Primum Non Nocere?" *Lancet* 344 (1994): 1487–1488.

14. Canby-Hagino, Edith, Javier Hernandez, Timothy Brand, and Ian Thompson, "Looking Back at PCPT: Looking Forward to New Paradigms in Prostate Cancer Screening and Research," *European Urology* 51 (2007): 27–33.

15. Chalmers, Thomas, "The Impact of Controlled Trials on the Practice of Medicine," *Mount Sinai Journal of Medicine* 41 (1974): 753–759.

16. Chlebowski, Rowan, Nananda Col, Eric Winer, Deborah Collyar, Steven Cummings, Victor Vogel III, Harold Burstein, Andrea Eisen, Isaac Lipkus, and David Pfister for the American Society of Clinical Oncology Breast Cancer Technology Assessment Working Group, *Journal of Clinical Oncology* 20 (2002): 3328–3343.

17. Colditz, Graham, "From Epidemiology to Cancer Prevention: Implications for the 21st Century," *Cancer Causes Control,* advance electronic publication, January 2007.

18. Crawford, E. David, William Fair, Gary Kelloff, Michael Lieber, Gary Miller, Peter Scardino, and Edward DeAntoni, "Chemoprevention of Prostate Cancer: Guidelines for Possible Intervention Strategies," *Journal of Cellular Biochemistry,* Supp. 16H (1992): 140–145.

19. Croker, Kara Smigel, Anne Ryan, Thuy Morzenti, Lynn Cave, Tamara Maze-Gallman, and Leslie Ford, "Delivering Prostate Cancer Prevention Messages to the Public: How the National Cancer Institute (NCI) Effectively Spread the Word About the Prostate Cancer Prevention Trial (PCPT) Results," *Urologic Oncology: Seminars and Original Investigations* 22 (2004): 369–376.

20. Demierre, Marie-France, Peter Higgins, Stephen Gruber, Ernest Hawk, and Scott Lippman, "Statins and Cancer Prevention," *Nature Reviews: Cancer* 5 (2005): 930–942.

21. Di Lorenzo, Giuseppe, Sisto Perdonà, Sabino de Placido, Massimo D'Armiento, Antonio Gallo, Rocco Damiano, Domenico Pingitore, Luigi Gallo, Marco de Sio, and Riccardo Autorino, "Gynecomastia and Breast Pain Induced by Adjuvant Therapy with Bicalutamide After Radical Prostatectomy in Patients with Prostate Cancer: The Role of Tamoxifen and Radiotherapy," *Journal of Urology* 174 (2005): 2197–2203.

22. Etzioni, Ruth, David Penson, Julie Legler, Dante de Tommaso, Rob Boer, Peter Gann, and Eric Feuer, "Overdiagnosis Due to Prostate-Specific Antigen Screening: Lessons from U.S. Prostate Cancer Incidence Trends," *Journal of the National Cancer Institute* 94 (2002): 981–990.

23. Fisher, Bernard, Joseph Costantino, D. Lawrence Wickerham, Carol Redmond, Maureen Kavanah, Walter Cronin, Victor Vogel, André Robidoux, Nikolay Dimitrov, James Atkins, Mary Daly, Samuel Wieand, Elizabeth Tan-Chiu, Leslie Ford, Norman Wolmark, and other National Surgical Adjuvant Breast and Bowel Project Investigators, "Tamoxifen for Prevention of Breast Cancer: Report of the National Surgical Adjuvant Breast and Bowel Project P-1 Study," *Journal of the National Cancer Institute* 90 (1998): 1371–1388.

24. Fleshner, Neil, and Girish Kulkarni, "Should Finasteride Be Used to Prevent Prostate Cancer?" *Current Treatment Options in Oncology* 7 (2006): 346–354.

25. Goetzl, Manlio and Jeffrey Holzbeierlein, "Finasteride as a Chemopreventive Agent in Prostate Cancer: Impact of the PCPT on Urologic Practice," *Nature Clinical Practice: Urology* 3 (2006): 422–429.

26. Goodman, Phyllis, Catherine Tangen, John Crowley, Susan Carlin, Anne Ryan, Charles Coltman, Jr., Leslie Ford, and Ian Thompson, "Implementation of the Prostate Cancer Prevention Trial (PCPT)," *Controlled Clinical Trials* 25 (2004): 203–222.

27. Goodman, Phyllis, Ian Thompson, Catherine Tangen, John Crowley, Leslie Ford, and Charles Coltman, Jr., "The Prostate Cancer Prevention Trial: Design, Biases, and Interpretation of Study Results," *Journal of Urology* 175 (2006): 2234–2242.

28. Goodwin, James, "Narcissus Drowned," *Lancet* 357 (2001): 1376.

29. Gorovitz, Samuel, and Alasdair MacIntyre, "Toward a Theory of Medical Fallibility," *Hastings Center Report* 5 (December 1975): 13–23.

30. Grover, Steven, Ilka Lowensteyn, David Hajek, John Trachtenberg, Louis Coupal, and Sylvie Marchand, "Do the Benefits of Finasteride

Outweigh the Risks in the Prostate Cancer Prevention Trial?" *Journal of Urology* 175 (2006): 934–938.

31. Hill, A. Bradford, "The Clinical Trial," *British Medical Bulletin* 7 (1951): 278–282.

32. Hill, M. J., "Prostate Cancer Screening" (Book Review), *European Journal of Cancer Prevention* 11 (2002): 313–314.

33. Huggins, Charles, "Endocrine-induced Regression of Cancers," Nobel Lecture, December 13, 1966.

34. Kaplan, Steven, "Medical Therapy for Asymptomatic Men with Benign Prostatic Hyperplasia: Primum Non Nocere," *Urology* 62 (2003): 784–785.

35. Karha, Juhana, and Eric Topol, "The Sad Story of Vioxx, and What We Should Learn from It," *Cleveland Clinic Journal of Medicine* 71 (2004): 933–939.

36. Klein, Eric, Catherine Tangen, Phyllis Goodman, Scott Lippman, and Ian Thompson, "Assessing Benefit and Risk in the Prevention of Prostate Cancer: The Prostate Cancer Prevention Trial Revisited," *Journal of Clinical Oncology* 23 (2005): 7460–7466.

37. Klein, Eric, "Can Prostate Cancer Be Prevented?" *Nature Clinical Practice: Urology* 2 (2005): 24–31.

38. Kulkarni, Girish, Rami Al-Azab, Gina Lockwood, Ants Toi, Andrew Evans, John Trachtenberg, Michael Jewett, Antonio Finelli, and Neil Fleshner, "Evidence for a Biopsy Derived Grade Artifact Among Larger Prostate Glands," *Journal of Urology* 175 (2006): 505–509.

39. *Lancet* editors, "When Primum Non Nocere Fails," *Lancet* 355 (2000): 2007.

40. Lippman, Scott, Steven Benner, and Waun Ki Hong, "Chemoprevention: Strategies for the Control of Cancer," *Cancer*, August 1, 1993: 984–990.

41. Lippman, Scott, and Waun Ki Hong, "Cancer Prevention Science and Practice," *Cancer Research* 62 (2002): 5119–5125.

42. Lippman, Scott, and J. Jack Lee, "Reducing the 'Risk' of Chemoprevention: Defining and Targeting High Risk—2005 AACR Cancer Research and Prevention Foundation Award Lecture," *Cancer Research*, March 15, 2006: 2893–2903.

43. Loefler, Imre, "Why the Hippocratic Ideals Are Dead," *British Medical Journal*, 15 June 15, 2002.

44. Lucia, M. Scott, Jonathan Epstein, Phyllis Goodman, Amy Darke, Victor Reuter, Francisco Civantos, Catherine Tangen, Howard Parnes, Scott Lippman, Francisco La Rosa, Michael Kattan, E. David Crawford, Leslie Ford, Charles Coltman, Jr., and Ian

Thompson, "Finasteride and High-Grade Cancer in the PCPT." Manuscript.

45. Marberger, M., J. Adolfsson, A. Borkowski, J. Fitzpatrick, D. Kirk, D. Prezioso, C. Rabaça, E. Solsona, and P. Teillac, "The Clinical Implications of the Prostate Cancer Prevention Trial," *BJU International* 92 (2003): 667–671.

46. Marks, Leonard, "5α-Reductase: History and Clinical Importance," *Reviews in Urology*, 2004, S11–S21.

47. Mellon, J. Kilian, "The Finasteride Prostate Cancer Prevention Trial (PCPT)—What Have We Learned?" *European Journal of Cancer* 41 (2005): 2016–2022.

48. Moinpour, Carol McMillen, Jonnae Atkinson, Sarah Moody Thomas, Sandra Millon Underwood, Carolyn Harvey, Jeanne Parzuchowski, Laura Lovato, Anne Ryan, M. Shannon Hill, Edward DeAntoni, Ellen Gritz, Ian Thompson, and Charles Coltman, Jr., "Minority Recruitment in the Prostate Cancer Prevention Trial," *Annals of Epidemiology* 10 (2000): S85–91.

49. Moyad, Mark, "Selenium and Vitamin E Supplements for Prostate Cancer: Evidence of Embellishment?" *Urology* 59; Supp. 4A (2002): 9–19.

50. Nelson, William, Angelo De Marzo, and William Isaacs, "Mechanisms of Disease: Prostate Cancer," *New England Journal of Medicine*, July 24, 2003: 366–381.

51. Paul, Yash, and Angus Dawson, "Some Ethical Issues Arising from Polio Eradication Programmes in India," *Bioethics* 19 (2005): 393–406.

52. Reuben, Adrian, "First Do No Harm," *Hepatology* 42 (2005): 1464–1470.

53. Rittmaster, Roger, "Finasteride." *New England Journal of Medicine*, January 13, 1994: 120–125.

54. Roehrborn, Claus, and Yair Lotan, "The Motion: Prevention of Prostate Cancer with a 5α-Reductase Inhibitor is Possible," *European Urology* 49 (2006): 396–400.

55. Scardino, Peter, "The Prevention of Prostate Cancer: The Dilemma Continues," *New England Journal of Medicine*, July 17, 2003: 297–299.

56. Shelton, James D., "The Harm of 'First, Do No Harm,'" *Journal of the American Medical Association*, December 6, 2000.

57. Smith, Cedric, "Origins and Uses of *Primum Non Nocere*—Above All, Do No Harm!" *Journal of Clinical Pharmacology* 45 (2005): 371–377.

58. Smith, Richard, "The Ethics of Ignorance," *Journal of Medical Ethics* 18 (1992): 117–18, 134.

59. Stoner, Elizabeth, "The Clinical Development of a 5α-Reductase Inhibitor, Finasteride," *Journal of Steroid Biochemistry* 37 (1990): 375–378.

60. Svatek, Robert, J. Jack Lee, Claus Roehrborn, Scott Lippman, and Yair Lotan, "The Cost of Prostate Cancer Chemoprevention: A Decision Model Analysis," *Cancer Epidemiology, Biomarkers and Prevention* 15 (2006): 1485–1489.

61. Tchou, Julia, Nanjiang Hou, Alfred Rademaker, V. Craig Jordan, and Monica Morrow, "Acceptance of Tamoxifen Chemoprevention by Physicians and Women at Risk," *Cancer* 100 (2004): 1800–1806.

62. Tercyak, Kenneth, Beth Peshkin, Barbara Brogan, Tiffani DeMarco, Marie Pennanen, Shawna Willey, Colette Magnant, Sarah Rogers, Claudine Isaacs, and Marc Schwartz, "Quality of Life After Contralateral Prophylactic Mastectomy in Newly Diagnosed High-Risk Breast Cancer Patients Who Underwent *BRCA1/2* Gene Testing," *Journal of Clinical Oncology* 25 (2007): 285–291.

63. Thompson, Ian, Donna Pauler Ankerst, Chen Chi, Phyllis Goodman, Catherine Tangen, M. Scott Lucia, Ziding Feng, Howard Parnes, and Charles Coltman, Jr., "Assessing Prostate Cancer Risk: Results from the Prostate Cancer Prevention Trial," *Journal of the National Cancer Institute* 98 (2006): 529–534.

64. Thompson, Ian, and M. Scott Lucia, "Diagnosing Prostate Cancer: Through a Glass, Darkly," *Journal of Urology* 175 (2006): 1598–1599.

65. Thompson, Ian, Chen Chi, Donna Pauler Ankerst, Phyllis Goodman, Catherine Tangen, Scott Lippman, M. Scott Lucia, Howard Parnes, and Charles Coltman, Jr., "Effect of Finasteride on the Sensitivity of PSA for Detecting Prostate Cancer," *Journal of the National Cancer Institute*, August 16, 2006: 1128–1133.

66. Thompson, Ian, M. Scott Lucia, Mary Redman, Amy Darke, Francisco La Rose, Howard Parnes, Scott Lippman, and Charles Coltman, "Finasteride Reduces the Risk of Developing Prostatic Intraepithelial Neoplasia." Manuscript.

67. Thompson, Ian, Phyllis Goodman, Catherine Tangen, M. Scott Lucia, Gary Miller, Leslie Ford, Michael Lieber, R. Duane Cespedes, James Atkins, Scott Lippman, Susie Carlin, Anne Ryan, Connie Szczepanek, John Crowley, and Charles Coltman, "The Influence of Finasteride on the Development of Prostate Cancer," *New England Journal of Medicine*, July 17, 2003: 215–223.

68. Trump, Donald, Joanne Waldstreicher, Geert Kolvenbag, Paul Wissel, and Blake Neubauer, "Androgen Antagonists: Potential Role in Prostate Cancer Prevention," *Urology* 57; Supp. 4A (2001): 64–67.

69. Unger, Joseph, Ian Thompson, Michael LeBlanc, John Crowley, Phyllis Goodman, Leslie Ford, and Charles Coltman, Jr., "Estimated Impact of the Prostate Cancer Prevention Trial on Population Mortality," *Cancer* 103 (2005): 1375–1380.

70. Unger, Joseph, Michael LeBlanc, Ian Thompson, and Charles Coltman, Jr., "The Person-Years Saved Model and Other Methodologies for Assessing the Population Impact of Cancer-Prevention Strategies," *Urologic Oncology: Seminars and Original Investigations* 22 (2004): 362–368.

71. Vaughan, E. Darracott, "Long-Term Experience with 5-α-Reductase Inhibitors," *Reviews in Urology* 5; Supp. 4 (2003): S28–S33.

72. Veurink, Marieke, Marlies Koster, and Lolkje de Jong-van den Berg, "The History of DES, Lessons to Be Learned," *Pharmacy World Science* 27 (2005): 139–143.

73. Walsh, Patrick, Urological Survey in *Journal of Urology* 174 (2005): 2221.

74. Wilt, Timothy, and Ian Thompson, "Clinically Localised Prostate Cancer," *British Medical Journal*, 25 November 25, 2006: 1102–1106.

75. Zeliadt, Steven, Ruth Etzioni, and Alan Kristal, "Estimated Impact of the Prostate Cancer Prevention Trial on Population Mortality," *Cancer* 1004 (2005): 1586–1587.

76. Zuger, Abigail, "A Big Study Yields Big Questions," *New England Journal of Medicine*, July 17, 2003: 213–214.

Index

A NOTE ON THE AUTHOR

Winner of the PEN Award for the Art of the Essay for his book *Seeds of Mortality*, Stewart Justman is professor of Liberal Studies at the University of Montana. Born in New York City, he studied at Columbia University and since the 1970s has lived in Missoula, Montana. He has also written *The Springs of Liberty*, *The Psychological Mystique*, *Literature and Human Equality*, and *Fool's Paradise*, a study of pop psychology. He is married with two children.